Innovate, Fund, Thrive

Entrepreneurs in life sciences face a unique set of challenges when raising funds. These encompass fundamental issues like navigating R&D risks and crafting a robust commercialization strategy, extending to more challenging hurdles like adeptly handling intellectual property issues, overseeing the regulatory development and approval processes, as well as coordinating extended and expensive phases of research and development.

The authors present this book in two parts. In the first part, the focus is on getting ready to meet potential partners and investors and includes topics such as identifying the information organizations and start-ups need, collecting and collating that information, and building a compelling story. The second part provides a deep dive into the investor's perspective, offering insights into how proposals are evaluated, along with an exclusive glimpse into the due diligence journey. Additionally, the book reveals "Eight Classic Mistakes Life Science Entrepreneurs Make", offering valuable insights and lessons gained from the author's experiences.

Although the book is divided into two parts, it maintains an interconnected approach. Both authors contribute insights based on their professional and personal experiences, offering feedback and perspectives throughout the book. Philippe provides his investor's perspective on the preparation steps, while Jean-François shares his insights as an experienced entrepreneur, coach, and mentor. This collaborative approach enhances the depth and practicality of the guidance provided.

Discover the art of preparing a compelling pitch, igniting excitement, and embarking on your path to fundraising success!

Innovate, Fund, Thrive

The Entrepreneur's Playbook to VC Fundraising in Life Sciences

Jean-François Denault and Philippe Tramoy

Routledge
Taylor & Francis Group

A PRODUCTIVITY PRESS BOOK

First published 2024
by Routledge
605 Third Avenue, New York, NY 10158

and by Routledge
4 Park Square, Milton Park, Abingdon, Oxon, OX14 4RN

Routledge is an imprint of the Taylor & Francis Group, an informa business

ISBN: 978-1-032-46501-2 (hbk)
ISBN: 978-1-032-46500-5 (pbk)
ISBN: 978-1-003-38197-6 (ebk)

DOI: 10.4324/9781003381976

Typeset in ITC Garamond
by KnowledgeWorks Global Ltd.

Contents

Preface

One late day, in May 2022, our two authors met for coffee.

Jean-François mentioned that VCs were difficult to pitch to and appeared arbitrary in their decision process. It was opaque from the outside. Philippe replied that many of the start-ups he met were often ill-prepared to present and lacked professionalism and preparation. After more back and forth, the idea for this book was born.

Our basic concept: creating a roadmap, including both our perspectives, to guide entrepreneurs in life sciences in their journey from idea to their first investment. To do this, we will share frameworks, checklists and ideas, while giving a look behind the scenes. Moreover, we will emphasize on the short term, which, after three years of COVID-19 pandemic, is certainly going to influence decision-making for the next few years.

Our complementary views and experiences were the glue of this book. While one of us would explain a topic, the other would weigh in and share his perspective on it. Hence, sprinkled throughout this book, you will find sections where "Philippe weighs in …" and "Jean-François weighs in ….".

The goal: make it interactive and share both sides of the coin (the entrepreneur and the investor) as much as possible. We have four goals with this book:

- **Increase transparency:** Make the VC decision process as transparent as possible and prepare entrepreneurs for their first financing experience. The investors' decision-making process is not very well understood by new entrepreneurs, and we believe that this book will provide a basic[1] understanding on how decisions are made, and what happens behind closed doors.
- **Increase credibility:** We want both parties to come to the table as credible stakeholders. To do this, entrepreneurs should know what

they need to share in their pitch. They should also understand how to document their proposal and position, and what type of support they need to prove their point.

■ **Increase trustworthiness:** We believe both parties should strive to be honest in their conversations. It is an underlying message woven throughout the book: getting caught in a lie or an "exaggeration" during the evaluation or due diligence is sure to derail the entire process, and as investing is a question of trust, breaking that bond reflects on the rest of the negotiation. In other words, we believe that entrepreneurs should address issues, rather than lying or hiding them. You have a competitor with strong technology? Focus on how your technology generates more interest from consumers. You do not have a full team yet? Focus on the team that is already present and explain your recruitment strategy. IP issues around your innovation? Show how you are working with your IP agent to develop a solution.

■ **Increase preparation:** Preparation is the key to success. As such, preparing your presentation, your business plan and your pitch should not be taken lightly. Think about the key topics, do pitch simulations with colleagues, practice answering questions. Being able to answer questions on the fly indicates preparation and knowledge and will be invaluable in making sure the next conversation occurs.

Overall, the objective of this handbook is to get life science start-ups investment ready. It is intended for the following audiences:

■ Emerging start-ups, developing their technology and looking to complement their go-to-market plan.
■ Young entrepreneurs, who need a roadmap to prepare for their fundraising activities.
■ Established coaches, mentors, incubators and accelerators who are looking for a "manual" for their clients or start-ups as they work with them to get them ready for fundraising.

 Before we jump into our content, I want to take a moment to thank my co-author, Philippe, for all the work in the last year as we wrote, debated, argued and corrected each other to make sure this book matched our vision of what we were looking to write. Authoring a book is an endeavor, and deciding to go the co-author route is a lot more challenging than I initially thought. Thanks Philippe, for all the ideas

and feedback, for being brutally honest and for making sure we create relevant content.

I also want to give a big shout-out to my clients, who I have had the opportunity to work with for so many years. For some clients, our relationships go back 20 years, for others, our journey is just beginning. From all of you, I learned so much and had the opportunity to work on so many different projects with you.

Finally, a big thank you to my wife, Corinne, and our kids, Jaz, and Chad, as well as my parents, Jean-Marc, and Daisy for their support throughout this project.

 I would like to acknowledge the extraordinary effort that Jf put into this book: he has taught me so many thing on writing over the last year. His "basics" and inputs every Friday evening were so much fun! You are right Jf, authoring a book is an effort, and deciding to go the co-author route is a lot harder than I first imagined. Thank you, Jean-François, for your patience and rigor, making sure we delivered relevant content.

We would not be able to get our work done without the continual support and vision of our editor, Kristine Mednansky at Taylor & Francis Group. Her staff copy-edited this with enormous skill and warmth and precision.

In France, Europe, but also worldwide, I thank students, entrepreneurs, colleagues, partners, collaborators, customers, investors, universities, research centers, BPIfrance, the European Investment Bank, Groupe BPCE, Natixis, the European Investment Fund, Seventure Partners, Sorbonne university, ENS Lyon, University of Burgundy, Paris Pantheon-Sorbonne University, Sorbonne Business School and HEC Paris school. I learned and still learn so much from you. Thank you!

Finally, I would like to thank with my infinite gratitude, support and love from my family - my parents Annie and Pierre (I miss you so much); my wife Elodie and our children, Leïa and Ylann. They all made me move forward directly or indirectly and this project would not have been possible without them.

It is our sincere hope that this book can help you in your innovation journey.

Note

1. Philippe believes I overuse the word "basic" in my writing style. I basically agree – Jf.

About the Authors

 Jean-François has been working with entrepreneurs in life sciences as a professional consultant for over 20 years. Through the years, he has worked with over 75 different clients in life sciences (including larger companies such as J&J, P&G, AbbVie, Denka Seiken and Novo Nordisk) as well as numerous innovative start-ups. His projects are focused on marketing strategy, commercialization, business modeling and business development. His clients are located all over the world, and he has completed projects with clients in over 25 different countries.

Jean-François also works (and volunteers) as a coach and mentor for several programs and accelerators, such as the HEC-Montreal University, NextAI, the European Institute of Technology and Innovation as well as ConnectInnov, a life science incubator located in Tunisia.

Jean-François has a graduate degree in Management Consulting, an executive MBA (specialized in the bio-industry) and a graduate degree in organizational communication. He is a member of the Editorial Boards of both the *Journal of Brand Strategy* and the *Journal of Digital & Social Media Marketing*. He has written a dozen articles for various publications and is the author of both the *Handbook of Market Research for Life Sciences* (Winner of the Outstanding Business Reference Sources Award (2018), awarded by the American Library Association) and the *Handbook of Marketing Strategy for Life Sciences*.

He is based in Montréal, Canada, and he can be reached through LinkedIn at https://www.linkedin.com/in/jfdenault/ or www.impacts.ca.

Philippe has been working in life sciences with wide-ranging responsibilities for over 20 years. He started research on a gene linked to leukemia and programmed cell death (apoptosis). He worked four years for a U.S.-based contract manufacturing organization (CMO), acquired by Rhodia (now Solvay), as a Business Development Manager for the EMEA region. Later, he founded and managed a market and business intelligence company with offices in France, Switzerland and Israel. He is a Partner for a Deep Tech fund at one of the European leaders in venture capital, financing start-up companies in the field of digital technologies and life sciences linked to academic and university institutions in France such as Sorbonne University, University of Paris IV, National Museum of Natural History, IRCAM, ENSCI, CNRS and Curie Institute. He is or has also been a board member of several Life Science companies.

Philippe also participates as a jury member or speaker for several programs and organizations, such as the MSC X-HEC Entrepreneur, iPHD launched by the French Government and Bpifrance and The Life Sciences Leadership School.

Philippe has a Master's in Biochemistry (University of Burgundy), a Master's in Immunology and Molecular Biology (Ecole Normale Supérieure Lyon), a Master's in Finance (IAE Paris-Sorbonne) and a Master's in Marketing (HEC Paris). He has authored articles for various publications and is the author of market research reports on topics such as the microalgae market, the European contract biomanufacturing market, cancer cytostatic compounds, Alzheimer's disease and private equity "the business of building business".

He is based in Paris, France, and he can be reached on LinkedIn at http://www.linkedin.com/in/philippetramoy.

Chapter 1

Overview of the Entrepreneur's Journey

  The entrepreneur's journey is an exciting one, which can take many different paths and roads. In both our collective journeys as entrepreneurs, mentors and investors, we have seen hundreds of different start-ups. Based on those experiences, we have seen a common framework emerge. It is this framework that we used to build our book, where we have framed five major steps in the entrepreneur's journey: The Idea, The Evaluation, The Development, The Pitch and The Breakthrough. We illustrate this journey in Figure 1.1.

Every Entrepreneurship journey starts with **The Idea**. It's the idea that drives the entrepreneur, the dream which he believes can change the world. In life sciences, the idea improves human health, and by extension, the existing healthcare system. It makes the healthcare system more efficient. It saves money, for the patient and everybody involved in the activity. It saves lives. At this step, the innovation is at its most basic form, moving from the problem to the solution. There might be some drawings, some early mock-ups, and a few texts or descriptions here and there. Maybe the entrepreneur had a few quick conversations to share the idea and get some feedback, but nothing really structured. This step is usually very short. For many would-be entrepreneurs, the project stops here: They either dismiss the idea and move on, or never really act on it, thinking about it for the rest of their lifetime. The most driven entrepreneurs are those who decide to act on the idea and move on to the next step.

DOI: 10.4324/9781003381976-1

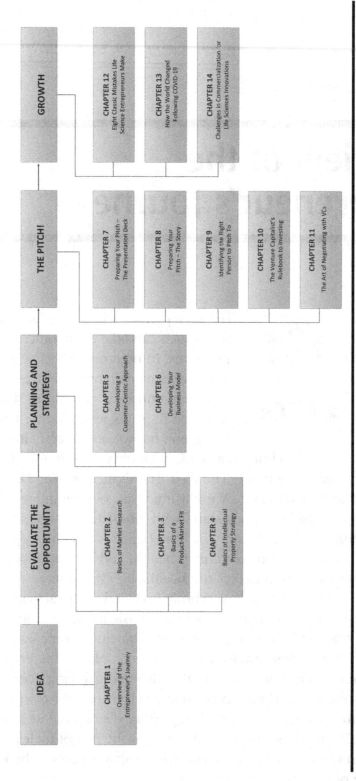

Figure 1.1 Overview of the entrepreneurship journey in relation to the content found in this book.

Those that move forward go onto **The Evaluation** step. This is where the entrepreneur becomes serious, sits down and starts validating if this is an idea worth investing in, both in terms of time and capital. For some entrepreneurs, it might start with a couple of evenings spent on the Internet, googling to see what's currently out there. For others, it might start by going to lunch with people they have identified as having experience in that space, to get a reaction to their idea. Their goal is also to learn what other companies are doing and to start gathering data. They are taking their first steps into market research.

When you are at this step, it's very important to keep asking yourself *"Is this idea worth investing into?"*. It's still early in the entrepreneur journey, and it's still possible to stop without falling into the Sunk Cost fallacy[1], or to pivot your idea if there is a better opportunity.

When an entrepreneur is in **The Evaluation** step, there are three main dimensions he should be focusing on:

1. *Focus on the technology dimension*: Is the idea technologically possible? Will it be cost effective? Do I have the resources to develop this technology? Is this technology better than existing technologies (from a technical perspective)? Are there any technical barriers to resolve?
2. *Focus on the market dimension*: Is there a sufficiently attractive market opportunity? Do I have a sustainable competitive advantage? Is this technology better than existing technologies (from a market perspective)? Are there any commercial barriers to resolve?
3. *Focus on the intellectual property dimension*: Is my technology impeding on existing patents? Is it patentable? Can I protect my innovation sufficiently to maintain my competitive advantage?

Three chapters in this book can help you in this early journey. Chapter 2 (Basics of Market Research in Life Science) and Chapter 3 (Basics of a Product Market-Fit) will be helpful for your market evaluation, while Chapter 4 (Basics of Intellectual Property) will be helpful to start understanding Intellectual Property. Jean-François's first book, *The Handbook of Market Research for Life Science Companies*, might also be useful in this step.

Following this step, the entrepreneur either acknowledges and recognizes the limits of his concept, returns to the **Idea phase**, or is convinced and moves on to the next step.

During **The Development**, the entrepreneur is hard at work planning and preparing his company. You will see that as time moves forward, the plans become more complex, and require more details. The first few fundraising efforts might be possible with only a certain level of preparation, but once you get to venture capital, all your ducks should be in a row.

During development, it's important to remember that your company is made of two complementary aspects: The strategic and the operational. In the strategic plan, you will be developing aspects such as your target customer, your commercialization model, your revenue model and your corporate strategy. We will be exploring these topics in Chapter 5 (Developing a customer-centric approach) and Chapter 6 (Developing your business model). Jean-François's second book, *The Handbook of Marketing Strategy for Life Science Companies*, might also be useful on this topic.

In parallel, some more operational aspects of your company will also have to be developed such as your research and development milestones, your regulatory plan, your budget and financing plan as well as your staffing plan.

Once development is complete, comes the hardest part of the journey: **The Pitch**. This is where the idea is presented to potential investors and/ or partners. It is a humbling experience, as you will most likely have to pitch your idea many times before you find the right investor and/or partner, willing to invest in your company.

To prepare you for this journey, we have devoted multiple chapters to this topic. We will be sharing the best practices on how to prepare your pitch deck (Chapter 7) as well as preparing for the storytelling aspect of your pitch (Chapter 8), while Philippe has developed a whole section of the book, describing how to identify the right person to pitch to (Chapter 9), what happens behind the scenes during the evaluation and due diligence process (Chapter 10), as well as taking you through the art of negotiation (Chapter 11). To also prepare you to this important step in your journey, we have also shared what we believe are eight of the biggest mistakes entrepreneurs make when pitching to VCs, which you will find in Chapter 12.

Following successful fundraising, starts the next step of the entrepreneur's journey, the **Breakthrough**, where the entrepreneur puts into action all the plans and strategies he has elaborated. He shifts from planning to execution. Often accompanied by the VCs, it's now his job to make sure the company succeeds, the innovation reaches the market and is sold to its target customers.

While we cannot go into every aspect of running a company, we have shared a few thoughts that might be useful such as how COVID-19 has changed the life sciences space (Chapter 13) and challenges in commercialization of life sciences innovations (Chapter 14).

As we wrap-up this section, we wanted to share a few additional insights on the journey of the entrepreneur.

First, in our experience, the cycle is not a simple unidirectional timeline. Hence, you might find yourself moving back to the evaluation step as you develop your technology and raise more funds. This is not a setback, but rather a natural flow of entrepreneurship. By the same token, creating an idea and taking it to market might generate new ideas, starting the cycle all over again.

Also, as you work through your progress cycle, it is very important to get feedback from some of the key stakeholders you meet. For example, pitching with a VC might not lead to an investment, but some comments during their evaluation might enable you to identify a gap in your presentation: You might have missed a major competitor in your competitive scan, been overly ambitious in your commercialization strategy, or might have overestimated some elements of your value proposition (while under-valuating others). The key here is to obtain feedback without showing all your cards. If you are too blunt, the VC might identify a key weakness in your current assessment, and you might not get the best deal possible. If you are too subtle, you might get answers that are not very useful. The key is to find the middle ground where you inquire for feedback while not showcasing your knowledge (and knowledge gaps). A fine challenge indeed.

Finally, it is important to consider that the cycle of an entrepreneur is quite different than the cycle of a VC. While entrepreneurs are always in a hurry, it is not unusual for a venture capital investment cycle to take anywhere from 12 to 18 months, from the first pitch to the negotiated term sheet. We will have the opportunity to over this even more in the chapters related to the investment cycle (Chapters 9 to 12).

Note

1. The sunk cost fallacy occurs when a decision-maker refuses to abandon an idea or project based on what has already been invested, rather than focusing on what is upcoming, especially if the idea is not good. In this situation, you will often hear sentences such as "But we've already invested $x" or "I've put in x hours into this project". The challenge is to recognize that these investments are gone, and you should focus on evaluating the project on its current merits and prospects.

Chapter 2

Basics of Market Research

 A few years ago, a client I was working with shared their pitch deck they were bringing to a conference so I could give them some feedback. After reviewing the slides, I was shocked by some of the major issues: market data sets that were over five years old, no market data generated from potential customers, market assumptions that vastly overestimated their market reach and so on. Any solid questions from an investor would have quickly unraveled his presentation. We had to rebuild the presentation to pitch something more realistic.

I want to emphasize this because the quality of your presentation is directly related to the quality of the information you have access to. The more precise the information, the more credibility you will convey. Market research is done for two target audiences. First, it is done for the start-up for its own needs: It helps it make optimal decisions and is done when the lack of information could cost the organization more than the cost to acquire the information. Second, market research is done to convince potential partners and investors, to demonstrate the validity of an organization's marketing decisions.

In the next few pages, we will be reviewing some basic market research concepts, followed by the basics of the more popular market research techniques (both primary and secondary research). More information on how to plan, execute your market research as well as how to analyze the data you gather is available in one of the other books of this series, *The Handbook of Market Research for Life Sciences*.[1]

 DOI: 10.4324/9781003381976-2

 **Philippe weighs in on the importance of market research:** One of the companies I advised was launching a new product in the field of CAR-T therapies. It was essential to do a market analysis in the preparation for the launch of the new treatment and to define the market access strategy in the United States, Canada and Europe. CAR-Ts are highly personalized technologies that require an important level of specialization to deliver to patients, and therefore not all healthcare institutions are ready to administer them. In this context, and to define the market access strategy for the new treatment, we needed to:

■ Understand and identify which centers would have the capacity to issue CAR-Ts.
■ Analyze bottlenecks for approval in relevant markets.
■ Accordingly, define the actions to be implemented to ensure access to your new CAR-T therapy.

After an investigation carried out in more than a hundred care centers in the United States, Canada and Europe, the analysis allowed the start-up to define the next key steps to follow to develop the skills of each center and promote rapid access to the market for your new treatment.

2.1 Basic Market Research Concepts

Market research is full of dichotomies. To fully understand the market research process, we will be going over some of the more important concepts: primary versus secondary research, quantitative versus qualitative research and inch-deep/miles wide versus deep dives research projects.

2.1.1 Primary and Secondary Market Research

Primary market research is a market research activity where the entrepreneur is actively engaged in research and is creating data. Simply put, it is the collection of data that did not exist until the researcher completed the market research activity. Some of the more popular tools are web surveys, in-depth interviews and focus groups. While costlier than secondary data, the data collected is tailored specifically to the entrepreneur's needs and will belong solely to him and his organization (meaning competitors will not have access to this specific information). When you engage in conversations with many potential clients, discuss their needs and collate that information, you are engaging in primary research.

Secondary market research is the collation of data that already exists. For example, it could be collected through a web search, or by aggregating news posts or articles. As such, articles (both scientific and business related) that you collate through your research, or the PowerPoint presentation of a competitor you downloaded from his website are all part of your secondary research. The entrepreneur collects and then transforms the data into something coherent and useful. While less costly, it is not always tailored to the entrepreneur's needs.

2.1.2 Quantitative and Qualitative Data

Data can be quantitative or qualitative in nature. *Quantitative data* refers to data that can be measured and numbered. "Counting" the number of potential clients for a product, calculating the number of products or doses of a drug a consumer uses each day or the average distance a patient is willing to travel to visit a specialized clinic are all different types of quantitative data. People in life sciences are usually quite familiar with quantitative data to quantify the technical aspects of products. Quantitative data in market research is often used to size markets and identify market segments and opportunities. Some data collection tools are better adapted for quantitative data. For example, surveys (both online and in-person) are usually the best way to generate an important quantity of quantitative data.

Meanwhile, *qualitative data* is data that is subjective and subject to interpretation. It can include anything from stories to discussions, observations or pictures. Some examples of qualitative data include personal reasons for preferences in consumer products, the impact of quality on customer purchasing patterns or the impact of packaging color on the purchasing decision. Data collected through interviews, focus groups and observation is usually of qualitative nature.

In general, quantitative data is perceived as "more real" and is easier to convey to an audience, but qualitative data is very useful to understand and contextualize the story behind the numbers.

To illustrate this, let me share a simple example is a project I did a few years back. My client, an Australian ad agency, was working on a promotional campaign for a big pharmaceutical company. To prepare the right promotional message, we had surveyed over three thousand consumers on their usage of painkillers. The quantitative data demonstrated what the preferred brands were, but without context, we could not understand why the number one preferred brand was the one my client perceived as cheapest and less efficient. It is only after analyzing the qualitative data in the survey related to why consumers made these purchasing decisions that we were able to uncover patterns in decision-making (the major categories of reasons people gave centered on topics such as the family's choice, health reasons, routine, price/sales and advertising). To deepen our understanding, we monitored spontaneous online discussions relative to the brands. We then found passion around the preferred product due to fewer perceived secondary effects, debates on the home brands versus branded products and patterns on how consumers perceived competing brands. We found that the brand that was preferred by the consumers, while perceived as less effective, was recommended more often since it caused less stomach aches, while the brand we expected would be more popular was perceived as being tougher on pain, yet harder on digestion.

2.1.3 *"Miles Wide" versus "Deep Dives" Research*

When approaching a market research project, a researcher will have to decide if he's going for a miles-wide approach or going for a "deep dive".

The *miles-wide/inch-deep* approach indicates the researcher is doing an overview of a segment, industry or competitive landscape. As such, the research project is deployed in a way to gather information on as many multiple data points as possible simultaneously. By its nature, it is very exploratory. By the end of the market research initiative, you will possess a high-level overview of the specific topic.

For example, a client of mine had designed a monitoring device with high precision and high durability. We could imagine several potential markets for it, and as such, the first market research product was designed around this requirement: we contacted and discussed with 20 different individuals, ranging from doctors, physiotherapists, nurses to professional sports athletes (runners, basketballers, baseballers, weightlifters) as well as coaches, teachers and sports enthusiasts. The objective: get a general understanding of multiple segments simultaneously and identifying the most promising ones for further analysis.

The *deep dive* approach implies focusing exclusively on a specific predefined topic. When doing this type of project, the top trends, competitors or issues have already been identified, and the market researcher deploys his energy to researching each of these topics in a very specialized manner. This means that rather than interviewing generalists, you will be interviewing key opinion leaders and consulting specialized resources on a specific theme.

Going back to our previous example, we identified a market segment which seemed especially interested in the monitoring device. Our deep dive then focused on practitioners of a specific sport (both professionals and amateurs), their coaches as well as stores and distributors specializing in this sport segment. At that time, the objective was to increase our understanding of pain points, as well as identify key elements such as branding and positioning opportunities, feature development priorities and pricing sensibility.

Market research projects will normally be a convergence of these two types of research, starting with the inch-deep miles-wide approach to gain a better appreciation of the overall opportunity, and concluding with a deep dive on the most interesting targets. Remember that it is important to align expectations with available resources: If you have a small budget and a limited time set, you might not be able to collect as much data as you want to. Try to find opportunities to maximize the resources you have: For example, you could identify a nurse who is a practitioner of the sport your device is made for, or a doctor who moonlights as a sports coach. These "dual" interview opportunities can generate useful data when on a limited budget.

2.1.4 Types of Market Research

The last topic to discuss is identifying the types of market research you will be engaging in. There are six types of market research, and each of them will have different objectives and approaches (Figure 2.1).

■ **Product research** is done by conversing with end-users. The objective of this type of research is to validate demand and identify key features.

Figure 2.1 Six common types of market research.

- **Market segmentation** is done by organizing focus groups and surveys. When doing this type of research, you are trying to identify underserved subsets of your target clients.
- **Pricing research** is done mostly through surveys and price testing. The objective is to identify the target price for your product as well as pricing elasticity. It is also possible to engage in secondary research to get a general sense of your competition's pricing structure.
- **Advertising testing** is done mostly through A/B testing, through focus groups and surveys. Here, the effectiveness of ads and messages is the main objective.
- **Brand awareness** is done mostly through primary research, although secondary research could yield interesting data points. The objective is to determine the main brands in the space, as well as the reach and potential of your own brand.
- **Satisfaction analysis** is once again done through a mix of primary and secondary research. Here, the company already has a product being sold, and the objective is to measure customer satisfaction.

2.2 Preparing Your Market Research Plan

Coherent and valuable market research follows a systemic approach. As such, like many other processes in life, it all starts with building a plan. While it is tempting to jump directly into "market research" and start collecting data, a detailed market research plan ensures that the data collected will be consistent and useful. As we will see, market research must be planned beforehand to ensure consistency in the data that is collected, as well as formalizing the endpoint. Preparing your market research plan breaks down into two distinct steps: identifying and formulating the problem, and then determining the research design.

The first step of the market research process is to identify and formulate the problem (or the opportunity). By formally defining the problem, the market researcher will focus his research effectively, ensuring all participants share the same vision and objectives for the project. As such, the problem identification step will usually involve discussions with decision-makers, a review of secondary data and conversations with key opinion leaders.

The topic of research is usually defined in a few words. For example, it could be to identify emerging market opportunities for a new technology, the size and segment of the current market, or developing a customer profile (around their specific needs, issues and problems). In keeping in line with our previous example, my client had developed an innovative technology, but needed to do market research for three objectives:

1. Identify the key segment that would be most interested in the technology
2. Demonstrate market interest in the technology
3. Learn the key attributes to further develop and emphasize for the target segment

Once you have identified your key market research objectives, the next step is to determine the research design. It is the approach that you will use to collect your data and guide you in choosing the specific methods you will use to collect the information you need. Some key questions you will answer at this step are:

■ Which method(s) will I use to collect data?
■ How will I connect with my target audience? Who will I need to connect with? How do I connect with them? Will I need to incentivize them? How?

- Which data collection tools will I use (telephone, in-person, internet)?
- What is my total budget (both monetary and timewise)?

The data collection phase splits into distinct phases. First, there is a design step where you will design your sampling plan and your tools, followed by the collection of data.

The sampling plan is the detailed framework of who will be contacted and what the expected sample size is. The sample size is crucial for the validity of the information you collect. It is important to define your sample size so you can know when you are done with data collection, rather than always second-guessing yourself.

Once the research question has been designed, and the methodology decided upon, it is time to design the research tool. For example, if you have decided to do interviews, you will have to design an interview guide. Building an interview guide ensures consistency between each interview, between each interviewer (for example, if multiple co-founders are conducting interviews independently), as well as being a useful tool to remember topics during the interviews themselves.

Ideally, you should test your research instrument before using it at large. Test your interview guide with a few potential participants: you might find that some questions are redundant, that some questions are missing, and some questions are misunderstood by your target audience. It is much more cost-effective to find this out at this stage than at the data analysis stage.

Follows the data collection step, which is often the most time-consuming step of the market research process. Once you have reached the endpoint of your data collection, it is time to start analyzing data.

2.3 Collecting Data – Primary Research

Primary data is information that is generated directly by the market researcher to answer his research question. For example, when an entrepreneur is doing an interview, managing an online survey, or performing client observations, he is gathering primary data.

Primary research is costlier to generate (in terms of both time and resources), but it is customized for the researcher's needs. If he has correctly designed his tools, he should be able to solve his research problems. Also, the data he collects is proprietary, so it belongs to the organization exclusively, becoming a competitive advantage.

2.3.1 Data Collection Methods

In the next few pages, we will be going over the basics of the three main market research data collection tools. These overviews are useful to understand and select the instrument that you need for your project. The methods we will be reviewing are in-depth interviews, focus groups and online surveys.

2.3.1.1 In-depth Interviews

In-depth interviews are interviews that are done one-on-one, between the researcher and the participant. These interviews consist of mostly open-ended questions. The objective is to explore the topic in a semi-formal format, gathering qualitative information. While in-depth interviews are costlier (in terms of time and money), they present several advantages over questionnaires and online surveys. For starters, interviews give more opportunities for the researcher to motivate the respondent to participate in a truthful manner, and to not abandon the interview halfway through. Also, interviews allow more flexibility in exploring secondary topics as opportunities in the interview emerge. The more exploratory the topic, the more useful the in-depth interview is, as it allows the researcher to change the order in questions, to prioritize some topics if time is short or to go more in-depth if the participant is resolved to have a rich narrative on a specialized topic.

There are several actions a researcher can do to enhance the interview process. First, *prepare your interview*: Conduct a quick due diligence on the interview target prior to your interview, to identify potential expertise and fields of specific interest. Also, *record the interview*: When you record the interview, you will be able to focus on the answers the participants are providing, as well as asking questions and exploring topics with your interviewee, rather than spending time taking notes. Finally, *listen to your interviewee*: You are trying to collect data, so that means the participant needs the opportunity to share his information. Be careful not to spoon-feed the answers that you are looking for. This leads to bad data collection, and does not reflect real market conditions. As I learned a long time ago, a good sign that an interview went well is that you talked very little and were able to learn a lot from your interviewee.

In-depth interviews in life sciences are a popular way to get the information you need, especially when dealing with topics which are sensitive in nature. It is a great way to speak to people confidentially and get their views on healthcare topics, such as their personal health and those of loved ones, their use of medical technologies and so on. It is also very interesting as the people you interview will often talk about competing products as they talk about your topic. Interviewing doctors and medical personnel can be especially challenging as these individuals are very solicited for their time. As such, you might have to set aside an important per diem to get them to participate. In projects I have done this in the past, it was not rare to have to pay up to $200/USD for a generalist doctor, and over $400/USD for a specialist.

2.3.1.2 *Focus Groups*

A focus group is a small group of individuals brought together to discuss a specific topic. The added value of a focus group (versus individual interviews) is that the interaction between individuals generates a wealth of information for the researcher. As such, focus groups are useful to gain a better understanding of what people are thinking, and why they are thinking it. It is also interesting as many participants will generate more information in a group setting where they feel safe, and they do not feel they are the sole focus of the interview. Finally, the information gathered in a focus group can be very useful to design a follow-up quantitative questionnaire or used in interpreting the information gathered from a quantitative research project.

The main issue with focus groups is issues you find in peer groups. As such, the presence of social pressure (the desire to conform to the group and agree), an individual's dominating personality in a group or the halo effect generated by key opinion leaders can all impact the quality of the interactions and the information generated. It is the role of the moderator to step in, to re-balance the focus group and ensure that it does not become biased. The other issue to remember is that while focus groups can be used to evaluate a group's feelings or views on a topic, it cannot be used as a final decision tool. It is exploratory in nature, not statistically valid, and the information you gather, while invaluable in interpreting existing data or setting up more research, cannot be an endgame.

Ideally, focus groups will have eight to ten participants, as the groups need to be sufficiently large to generate dynamic conversations, but not too large as to leave some participants out or become difficult to manage for the moderator (as parallel discussions start to emerge). About 45 to 75 minutes is the ideal period for a focus group: too short, and you will not generate any meaningful insight, while too long, you risk participant fatigue where they quickly agree in the hopes the focus group will end. Three to four focus groups are usually necessary to fully explore a topic: after that, you may find the same information being repeated.

In life sciences, focus groups are especially useful to interact with end-users (doctors, nurses, laboratory technicians) as well as getting feedback from patients. As such, they can generate valuable information relative to marketing, branding, competitors and product issues. Nonetheless, some researchers have found that focus groups are not an ideal environment for eliciting emotional information from physicians, because physicians' self-perceptions and the image they want to project to others are typically one of a rational (not emotional) decision-makers.[2] Also, it can be especially challenging to reach and recruit specific participants.

2.3.1.3 Online Surveys

The use of online surveys has grown immensely in popularity, almost completely replacing traditional mail-in surveys. They are cost-effective, simple to use and if done properly, can reach a wide range of populations, allowing the participants to complete the survey quickly on their own time with little effort. Most online surveys today are done using a web-based survey tool, but some organizations will use basic programs such as Word and Excel for smaller samples.

There are considerable challenges in getting participants for a life science survey, particularly with clinicians. In a 2021 meta-study done on articles found in the National Library of Medicine, researchers found that US healthcare professionals had an average response rate of 48.0%, compared to patients posting an average 64.2% rate. Healthcare providers in France did little better, where healthcare professionals had an average response rate of 47.3%, compared to patients' 72% response rate.[3] Lack of time and survey burden were the most common reasons for not participating.[4]

Another study found that general practitioner survey rates could be increased with incentives (larger and upfront, if possible), peers pre-contacting targets by phone, personalized packages and sending the survey on a Friday.[5]

There are several things to remember when building a web survey, and to increase your response rate. First, *keep the survey short and simple*. This helps to reduce user attrition. Second, be *straightforward about the time to answer.*

Announce upfront the length of the survey and, if possible, use a progress bar on top of the survey to keep participants engaged, Finally, *optimize your survey for mobile devices*: Use a survey platform that will optimize your survey for mobile platforms: more people use mobile devices to do mundane tasks while waiting, and if your survey does not properly display on a mobile device, participants might simply drop-out and move on.

2.4 Secondary Research

Secondary research is the collection and collation of information that is published and publicly available. Also called desk research, it is frequently done to explore a topic before engaging in more expensive primary research, or to quickly gain a summary understanding of a topic without engaging too many resources.

The advantages of doing secondary research (versus primary research) are numerous. First, secondary research is less expensive than primary research. Also, it is faster to complete as it does not depend on third parties (such as recruited participants, organized focus groups or enrolled survey participants) to obtain the information, and the sample size of third-party research reports will often be quite considerable. Furthermore, due to scope and reputation, third-party information often has more perceived authority and impartiality than "in-house" research. An assessment such as "BCC Research estimates that the global advanced drug delivery market should grow from $178.8 billion in 2015 to $227.3 billion by 2020, with a compound annual growth rate (CAGR) of 4.9%"[6] has more credibility than most in-house estimates. Finally, as mentioned earlier, secondary research is very useful to orient and define primary research: a researcher will often start a market research project by doing a quick market review to identify some of the main trends and concerns before diving into primary research.

 **Philippe weighs in on the importance of understanding the market:** It is essential to invest in understanding the market, especially key subjective elements. What are the barriers to entry? What are the competing services or products or technologies? What is the R&D activity of market players?

A market is like a living entity. It is constantly evolving. Between the moment "m" when you analyze it, you photograph it and the moment when you will be able to address it, all the actors will be in motion, same as you. To illustrate, imagine your company is developing a new drug product in a therapeutic area with no solution. Six years later, a new technology (immunotherapy for example) addresses the market with a very relevant solution. The window of opportunity closes. Either your cash allows you to pivot, or the venture is doomed. The market has evolved. It is therefore essential to combine primary and secondary information, and to always be particularly alert.

There are some disadvantages to secondary research. First, the information is not always personalized to an organization's requirements, and it is quite difficult to find data for emerging fields: reports on nanomedicine are plentiful, but developing a specific application merging nanomedicine and information technology means that secondary research will be quite scarce, and that the organization will either need to a) extrapolate from generalist research or b) conduct primary research. Also, the data might be outdated, limiting its usefulness. Finally, most of the time, the original data used in secondary data is unverifiable: It is quite difficult to spot errors in data collection or dispute the way data was analyzed in a consolidated report.

There are two types of secondary data: external and internal secondary data. External secondary data is information gathered from outside the organization. This includes anything from government statistics to media sources. Internal secondary data is data that the organization is generating itself. It could be data collected from customers' feedback, accounting and sales records or employee experiences. While it is possible to do secondary research using non-internet sources, the bulk of our suggestions are related to this medium for both the ease of use, convenience as well as cost.

Finally, there is a distinction between active and passive secondary research. Active secondary research takes place when the researcher is actively searching for information, while passive secondary research is the use of tools and software to automate data collection.

2.4.1 Active Secondary Research

Active secondary research takes place when the market researcher is dynamically searching for information. In the next pages, we will be going over the most popular and pertinent sources of data available. All the suggested sources are web-based.

While there are some ethical considerations for what information the researcher can use, a simple rule of thumb is that any information made available to the public is fair game for collection and review. If the method used to obtain the information is not commonly available to the public (such as using a former employee's password to access a restricted website area), then it is not only unethical, but most definitely illegal.

2.4.1.1 Popular Sources of Data Online

- **Government data:** Government agencies generate large bodies of information which can be used by entrepreneurs. Most of this information is free to use and can be useful at the start of a research project. Government data is usually statistical in nature and is very useful when building marketing models or trying to understand the nature of a market. Some key databases include the National Center for Health Statistics (https://www.cdc.gov/nchs/index.htm), the World Bank Open Data (https://data.worldbank.org/) and Eurostat (https://ec.europa.eu/eurostat).
- **Public company data:** Companies publish a lot of useful information online. Often, companies will do market research, and will publish results in their corporate documentation. As such, reviewing this publicly available data is another useful way to start your research effort. Some of the documents that can be reviewed online include Annual reports, company pitch decks, press releases and video product presentations. The most useful databases are found below in Table 2.1.
- **Print media:** Print media sources are published on a regular schedule by specialized companies. These include magazines and newspapers and include both their printed/physical format as well as their internet counterparts. There are several trade magazines that are published on a regular basis which are of use to market researchers. Most of them are free, and access to their archives is public most of the time, although

Table 2.1 Table of US, Canadian and European Databases of Public Company Data

Country	Website
United States	EDGAR (https://www.sec.gov/edgar/search-and-access)
Canada	SEDAR (https://www.sedar.com/)
United Kingdom	Gov.UK (https://www.gov.uk/get-information-about-a-company)
France	AMF (https://www.amf-france.org/fr)

some do monetize their archives. Popular ones include Genetic Engineering and Biotechnology News (https://www.genengnews.com/), Fierce Biotech (https://www.fiercebiotech.com/) and MD+DI (https://www.mddionline.com/).

- **Social networks:** For this book, "social networks" are defined as dedicated websites or applications where users with a common objective aggregate and participate in discussions. These discussions create a network of social interactions, as users share messages, comments, information, opinions, experiences and more. Social networks are a useful way to get the customer's pulse on a topic. Analyzing the comments found on social network pages allows the researcher to understand how people view a brand, or their perception of a specific topic. It is also possible to use social networks to recruit participants for surveys and focus groups, as well as interacting directly with them and engage in one-to-one conversations. Beyond the usual suspects (Facebook and LinkedIn), I have found that Reddit is often a good resource to find leads on potential competitors, market trends and pricing information.

- **Trade and industry groups:** Trade and industry groups are organizations representing multiple firms (private companies, government agencies, universities, consultants) in a common commercial activity sector. Some of the larger associations produce (or sponsor) reports relevant to their industry, which can be of use during market research. One of the advantages of the reports produced by these organizations is that they are third-party reports and (relatively) unbiased, but the reports are not always freely available, and are sometimes reserved for members. Some of the main trade groups you should be keeping an eye on (depending on your specific industry sector) include the Biotechnology Innovation

Organization (https://www.bio.org/), PhRMA (https://www.phrma.org/), the American Hospital Association (https://www.aha.org/) and the Medical Device Manufacturers Association (https://www.medicaldevices.org/).

Philippe weighs in on secondary resources value: A company's internal data, such as sales and marketing records, customer account information, product purchasing and usage data are typical secondary data sources of interest. It is always extremely interesting to delve into the annual reports of market players to fully understand the market segmentations, issues, market sizes and trends. The parts dedicated to market analysis are often very rich in information and particularly detailed, especially in listed companies. For example, in Merck's 2022 annual report,[7] you will learn that the ex-COVID life science industry is worth ~€200 billion and that Merck Group has 5% of market share. Also, a very good "free" source of information is the company's initial public offering (IPO) document. The evangelism part of the market especially for deep tech companies in the life sciences will be well organized and very detailed. Do not reinvent the wheel and tap into the countless sources of information available. General market studies are good tools for understanding the market and its structure, but the projections are often optimistic and the CAGRs (compound annual growth rate) displayed are often overestimated. In all cases, studies carried out by teams with a strong connection of the sector and industrial experience are to be preferred.

2.4.1.2 Using Search Engines to Look for Information

Searching through the internet using a web research engine is usually the first step people take when engaging in secondary research. Here are a few tips and tricks to make the research effort more efficient

- *Go beyond Google:* If you are not finding the information you need, you can try using another search engine to obtain different search results. Some of the interesting alternative search engines include
  - Bing (https://www.bing.com/), which is reported to have a better Video Search Option;

- Board reader (https://boardreader.com/), which specializes in user points of view by searching through forums, message boards and Reddit.
- Slide Share (https://www.slideshare.net/), a cornucopia of PowerPoint presentations, slide decks and webinars from past conferences.

■ *Look for corporate web DNA*: I originally found this technique referenced by Leonard Fuld in "The Secret Language of Competitive Intelligence". It is based on the concept that every organization develops their own brand of corporate speak, or pattern. It is akin to corporate web DNA. He defines it as *"a unique pattern of words and phrases that form the substance of a company's website, its press releases and its advertisements"*.[8] As such, if the researcher can identify a group of unique words or jargon as potential corporate DNA, he can then proceed to researching the web using the above-mentioned terminology, grouped between two sets of quotations. As an example, using Medtronic's *"life-transforming technology"* slogan to search the web brings up a series of white papers (old and new), job offers (both current and expired) as well as customer testimonials.

■ *Look to the past*: Sometimes, a market researcher will be looking for something specific, but will conclude that the information is no longer available online. For example, it might be an old press release that a competitor has pulled from their website, information on a previous partnership that has been quietly ended or specifications on discontinued products. In these cases, the website Archive.Org (an Internet Archive non-profit digital library offering free universal access to all) is especially useful. Archived in their public database are historical web snapshots of the company's website, which can include pages, attachments and more. While a researcher might not have access to each version of a company's website, there are often several snapshots taken throughout the year, enabling the researcher to identify key information that has been removed online.

2.4.2 Passive Secondary Research

Automated internet research tools are a boon to market researchers, as they automatically monitor and report on specific information topics. We will be going through some of the most interesting tools researchers can use to automate their market research.

- **Google Alerts:** Google Alerts is a service offered by Google. It is useful to automatically monitor a topic by setting up a search alert. Once setup, the Google Alert sends search-engine search results by email to entrepreneurs as they occur, or on a predetermined basis in the form of a digest (once a day or once a week), at a predetermined time. To create a Google Alert, the user only needs to go to the website https://www.google.com/alerts, type in their topic of interest and customize the information feed requested (frequency, where the information will be collected from, the number of results wanted each period, and which email will be receiving the information). It is possible to edit an alert if needed or delete the alert if it is no longer needed.

- **RSS feeds:** RSS (Rich Site Summary) feeds are a simple method to aggregate data generated by specialized websites, and efficiently supply researchers with up-to-date information on specific topics. The advantages of using RSS feeds for research and continued monitoring are multiple. First RSS feeds save time: you can quickly subscribe to the feeds you are interested in, and quickly scan aggregated data without having to visit every single website every time. Also, as RSS feeds update themselves automatically, you get information as it becomes available. While the popularity of RSS feeds has diminished in the last ten years, over 225 million internet users still use RSS feeds to keep tab on the internet, and at the time this book was written, many useful sites such as Reddit, CNN and Slashdot, for example, still had RSS Feed capability.

  Three types of RSS Readers exist. There are web/mobile-based readers (which you access through your web browser or mobile device) such as Feedly (https://feedly.com/) or Inoreader (https://www.inoreader.com/), client-based readers which you download and install on your computer such as RSSOwl (https://www.rssowl.org/) and those that integrate into your web browser (most popular web browser offer this option, or allow you to install add-ons that allow you to monitor RSS feeds).

- **Social media tracking:** There are several tools and search engines that are specifically designed to monitor social media. These tools search through popular Social Media websites such as LinkedIn, Facebook and user-generated content such as blogs and comments to generate reports which can be used to identify underlying consumer trends. One of these popular tools in this space is Keyhole (http://keyhole.co/). This monitoring tool keeps an eye on keywords and

#hashtags across X (formerly called Twitter) and Instagram and can be useful to quickly identify popular public key opinion leaders on a topic, which can then be engaged further for market research. It can also be used to identify geographic trends and estimate overall consumer sentiment. Other popular options include Mentionlytics (https://www.mentionlytics.com/), Hootsuite (https://www.hootsuite.com/) and Sprout Social (https://sproutsocial.com/).

2.4.3 Internal Secondary Data

Internal secondary data exists and is stored inside the organization. It is available exclusively to the organization and is usually generated and collected during normal business activities. Internal sources of data should always be investigated first because they are usually the quickest, mostly inexpensive and most convenient source of information available.

Sources of internal secondary data include

- *Sales data*: If the organization is commercializing its product, it has access to an invaluable resource, its own sales data. This data is usually collected by the organization and organized in a way that is useful and extractable by a market researcher. Some of the data the researcher can look at includes sales invoices, sales inquiries, quotations, returns and sales forces business development sheets. From this information, territory trends, customer type, pricing and elasticity, packaging and bundling impact can be inferred. This data can be useful to identify the most profitable customer groups, and which ones to target in the future.
- *Financial data*: All functioning organizations have accounting and financial data. This can include cost of production, storing and distributing its products. It can also include data on research and development costs and burn rate and can be used to calculate valuable ratios.
- *Internal expertise*: Mid-sized organizations will often have inside expertise in the form of its personnel that have been with the organizations for a while. These individuals can be tapped and interviewed to get more information on past initiatives, products, lost customers or any other topics. Often referred to as the organizational memory, they often have a wealth of undocumented knowledge that can be harvested. They might be aware of internally produced reports that might be of use, past projects or failed product initiatives.

Some of the weak points of internal data are inaccuracy, as it might be dated or there might be issues with the way it was collected. Also, while most of the time data can be ported from internal systems to market research data analysis tools, some legacy systems might make data conversion especially challenging. Finally, there are confidentiality issues: some companies employing third-party researchers might hesitate to "open the books". In this case, it could be possible to share either consolidated data, or limited data sets.

2.5 A Few Words on Ethics and Market Research

We presented and discussed some of the tools an entrepreneur can use to collect data. But there are a few things to remember relative to ethics before engaging in market research.

First, it's very important to be honest when you collect data, *identifying yourself and describing why you are collecting the data.* Misrepresentation is a huge issue in data collection, and it is tempting to disguise your identity to ease the data gathering. For example, some researchers will pretend to be a potentially interested client or posing as a student gathering information for a school project. This is clearly unethical. Instead, if you are having issues connecting with people to interview, efforts should be spent identifying targets which have the information and are more likely to want to share it such as academics, technology vendors, advertising agencies and journalists, for example. Furthermore, I have found that new entrepreneurs working on a new product and technology usually get a lot of inroad and availability, especially if they are young entrepreneurs designing their first product. If this applies to your profile, you should lead with that as you connect with the people you want to interview.

Also, *be neutral when asking questions.* It is very easy for a market researcher to influence the participant's response. Asking leading questions can cause a participant to answer in a specific way. Even agreeing with a participant rather than impartially acknowledging their answer can influence the participant. Remember that leading participants *might get you the answer you want to hear, but will not necessarily reflect the real market's appreciation of your product:* Wouldn't you rather find out the real market interest for your product during the market research phase, rather than shaping market research to fit your preconceived ideas, and then failing during commercialization? *While market research shapes our vision of the market, it does change the true nature of the market.*

Furthermore, *respect the confidentiality of the participants.* If you have given them the assurance that you will protect their responses, be ready to do so. If you believe you cannot assure the participants' confidentiality, or if you do not intend to (by sharing raw data with other stakeholders), be upfront with participants so they can have the option of opting out. Also, if during analysis, you identify a quotation you feel would be strengthened by attributing it to the specific person you interviewed, you should go back to that individual and get authorization to quote them. It only takes a quick email and might even lead to more insight.

Finally, *primary market research is not a commercialization activity.* Engaging participants in market research, and then trying to sell a product midway during data collection undermines market research. In some cases, a client might express interest in a product you are researching. When this happens, my approach is to ask the participant "You seem to have some interest in the product/service we are discussing. Would you like me to refer you directly to the Company as an interested party? Do you accept that your coordinates be shared directly to the appropriate person?" As such, the participant is authorizing you to share his information and his interest. You are serving both parties, and with consent, are relieved of other obligations you might have (such as keeping the data anonymous). And best of all, a successful interview (that didn't turn into a sales pitch) means you have successfully validated market interest, and you might be able to tap that person later to ask more questions, do a deep dive or discuss a completely different project.

Notes

1. Denault, JF. 2017. The Handbook of Market Research for Life Sciences, Productivity Press; 1st edition, 226 pages.
2. Kelly, D, and Rupert, E. 2009. Professional emotions and persuasion: Tapping non-rational drivers in health-care market research. *Journal of Medical Marketing: Device, Diagnostic and Pharmaceutical Marketing*, 3–9.
3. Meyer, VM, Benjamens, S, Moumni, ME, Lange, JFM, and Pol. RA. 2022. Global overview of response rates in patient and health care professional surveys in surgery: A systematic review. *Annals of Surgery*, 275(1), e75–e81. https://doi.org/10.1097/SLA.0000000000004078
4. Cunningham, CT, Quan, H, Hemmelgarn, B, et al. 2015. Exploring physician specialist response rates to web-based surveys. *BMC Medical Research Methodology.* doi:10.1186/s12874-015-0016-z.

5. Pit SW, Vo T, Pyakurel S. 2014. The effectiveness of recruitment strategies on general practitioner's survey response rates – a systematic review. *BMC Medical Research Methodology.* doi:10.1186/1471-2288-14-76.
6. Wadha H. 2016. Global Markets and Technologies for Advanced Drug Delivery Systems. http://www.bccresearch.com/market-research/pharmaceuticals/advanced-drug-delivery-systems-tech-markets-report-phm006k.html (accessed June 14th, 2023)
7. Merck, KGaA. 2022. Annual Report. https://www.emdgroup.com/investors/reports-and-financials/earnings-materials/2022-q4/us/2022-Q4-Report-NA.pdf (accessed June 14th, 2023)
8. Fuld, L. 2006. The secret language of competitive intelligence. *Crown Business.*

Chapter 3

Basics of a Product-Market Fit

 For an entrepreneur, demonstrating that their innovation is right for the market and will generate interest is a key step to raising funds. As such, it is important to have a framework that lets you validate if your product is a good fit for the market. One of the frameworks that has rapidly gaining popularity is the Product-Market Fit (PMF).

PMF is used to validate that your product's value proposition will generate actual customer interest to purchase your product, which will translate into real sales. This step usually follows the Problem-Solution Fit (PSF), which is used to identify the best solution to solve a problem. The difference between the two is quite noticeable: if you did a PSF, you might have developed a product that is the perfect solution to a problem, but you might not find anybody willing to buy it (due to cost, complexity and so on). As such, measuring PMF ensures that you have demonstrated the **market intent** in purchasing your product.

 **Philippe weighs in on the importance of PMF:** A startup is a high-risk venture. The level of risk decreases as the venture progresses toward what is called "proof of concept". Analytical analysis of risk factors will include the maturity of the management team, its financial strength, and the decisive factor is the degree of responsiveness that the company's product or service receives from the target market - PMF.

DOI: 10.4324/9781003381976-3

Before a startup rushes to reach the Proof of Concept (POC) stage, it should first tackle PMF. So PMF is indeed important. What is key, and more particularly for "DeepTech" companies that develop solutions or products in the life science sector, is to be able to create proof of concept opportunities with industry players in the sector. A proof of concept with manufacturers or a group of users or patients is a real milestone in value creation. In therapeutic product development, it is a question of having a preclinical trial or clinical trial validated by a health agency. For the others, having a "knower" who validates your solution, or your product is an important element.

3.1 Getting Your Data

We will be exploring three main methodologies to gather data and measure your PMF: conversations with potential clients, evaluation of online reviews and online surveys.

3.1.1 Conducting Conversations

The easiest methodology to measure PMF is engaging in conversations with potential clients, and evaluating their interest in the product or service you are developing. In life sciences, this might mean contacting patients, hospital purchasers as well as healthcare personnel (such as doctors and nurses). The idea is that by interacting with your future clients, you can identify something they react positively to, and then focus on your product or service on that key element. You can also reconfigure and adapt your product, as well as validating some of the features you had hypothesized might be of interest to your clients. Finally, you can quickly test your hypothesis around pricing as well as competing technologies.

If, after your investigation, your current product does not have a market fit, your efforts will have to be dedicated on retooling your product. This might mean changing the target market, changing the product or both.

There are multiple ways you can move forward in this space. You could contact individuals directly, request interviews, meet them at conferences and so on. Developing a deep understanding of potential clients is essential, so you can develop a significant value proposition. While more is better, contacting at least ten potential end-users is necessary to ensure you have a good sample.

As we mentioned earlier, be neutral when asking questions. It is quite easy for a market researcher to influence the participant's response. Asking leading questions can cause a participant to answer in a specific way. Even agreeing with a participant rather than impartially acknowledging their answer can influence the participant. Remember that leading participants might get you the answer you want to hear, but will not necessarily reflect the real market's appreciation of your product: wouldn't you rather find out the real market conditions for your product during the market research phase, rather than shaping market research to fit your preconceived ideas, and then failing during commercialization? *While market research shapes our vision of the market, it does not change the true nature of the market.*

A few years ago, I was hired to evaluate PMF for a technology developed for blind individuals. While the technology was promising, its current application proposition was limited, leading to a very niche (an unprofitable) market. Interviews with nurses, teachers, non-profit personnel as well as blind individuals enabled us to not only measure PMF, but also identify two new applications with a lot more market potential. This realignment enabled the client to plan new product lines which would generate profits, driving growth for the product.

3.1.2 Collecting Online Reviews

An alternative to engaging consumers directly is to evaluate online reviews of related products. Social media has made accessing massive amounts of text and feedback a lot easier, but most of the data that is generated is qualitative in nature. Nonetheless, it is possible to extrapolate user sentiment by using simple techniques such as word frequency, sentiment analysis and paired words.

- *Word frequency* is the analysis of reviews by identifying which are the most frequently used words that come up in each category. As such, word density is used to indicate the emergence of trends.
- *Sentiment analysis* is completed by using an online tool that uses a database to match certain words to sentiments. This is used mainly when you have a large database of qualitative data to analyze (such as a survey with many open-ended responses).

■ *Relationship between words*: Which are the most common pairs of words in the category? Exploring which words usually appear together in online reviews provides insight into how consumers associate concepts with one another.

While not as rich as talking to end-users, analyzing existing conversations can have several advantages. First, it can help you identify patterns and problems you had not identified yourself: it can be difficult to inquire about issues you have yet to find out exist. Second, it can help you identify gaps and opportunities in competing products: these gaps might become opportunities for you as you develop your product. Finally, it is less expensive than setting up a series of interviews with key stakeholders.

A project I did a few years ago included the evaluation of online reviews. The objective was to not only explore PMF, but also explore competitive gaps. As such, we gathered a good database of customer reviews and comments from the usual public forums (Amazon, Reddit and so on). Using word frequency tools, we found that the competing products had the words "doubt", "problem" and "lack of confidence" in more than one comment. This led my client to focus development on attributes where he could effectively demonstrate trust and confidence.

If you want to learn more about this topic, an article by Myriam Alzate called "What are consumers saying online about your products? Mining the text of online reviews to uncover hidden features" is a particularly valuable resource on the topic.[1]

3.1.3 *Organizing Online Surveys*

When preparing a product that will be purchased and used by the public at large, it is possible to organize and purchase a survey of "consumers". Using a third-party service such as Survey Monkey or Pollfish, an entrepreneur can quickly deploy a survey with a dozen or so basic questions, and then survey hundreds of potential clients for a low cost.

This will generate valuable market data the entrepreneur can then use to evaluate his product. Of note, the data will be a good indication, but should not be taken as gospel. Also, careful work must be put into how questions are formulated. For example, asking "Would you purchase this product?" and then getting a 90% interest from users is not an indication of overwhelming product interest: in my experience, users answering these paid surveys are usually overly positive, and will most likely answer in a positive fashion as to increase chances of being recruited and being able to do future surveys. A more subtle approach would be to ask, "What is the price you would be willing to pay for this product?" and then removing outlier answers.

Personally, I have used this method to get results, especially when combined with interviews and conversations. I have found that data points from one tool, rebound well and lead to interesting conversations (the same way that data from interviews enable great questions for surveys). In a recent project, I was conducting interviews to confirm PMF for an innovative exercise monitoring device. As conversations with physiotherapists and doctors were generating great data, there seemed to be an immense potential for amateur athletes as well. As such, we took some time to build a questionnaire around this topic, and then conducted a survey of 400 amateur athletes. This data, once compiled and analyzed, was then used to build a specific interview guide for the semi-professional coaches. This interaction between both tools delivered the best results for confirming market fit.

One last word of advice: if you do decide to conduct your own survey, I heartily suggest you consult Chapter 2 as well as my first book, *The Handbook of Market Research for Life Sciences*. In this, I go over extensively the techniques and tools you need to build a solid and valid online survey, which will ensure the data you collect will be of use.

3.2 Evaluating PMF

So how do you know you have a good market fit? One interesting framework to evaluate your PMF was developed by Adam Fisher, of Bessemer Venture Partners. He uses it to quickly evaluate companies in his portfolio as well as for founders to reflect on their own PMF journey.[2]

On the x-axis, we measure the depth of customer engagement, which we uncover by actively talking to the potential customers, understand their needs and current concerns. Entrepreneurs who are either out of touch, or talking to a limited number of stakeholders already convinced of their technology might not be challenged and as such might believe they have a good grasp of their customer. In reality, they are stuck in what he calls an echo chamber.

It can be challenging to balance between integrating customer feedback and sticking with your vision. Obviously, you do not want to shift your product to every feedback, but then again, it can be difficult to decide which feedback to listen to. Nonetheless, there are some things you can do. First, take care to notice patterns in feedback. A lone potential customer might not be a great indication, but if you are able to guess what the customer is about to say before they end their sentence, it might be time to seriously consider the feedback. I had a start-up I was working with, and I was sharing some feedback on their product, to which the founders responded "Yeah, everybody says that, because they don't understand what we're trying to do". In this case, it might be the contrary that is occurring, where the founders are not listening to the feedback, and refusing to adjust their product.

Other things you could do is dig deeper with the persons sharing feedback (You mentioned feature X is essential; can you tell me why? How would you use it? Do competitors already do it? How does it help you accomplish your work?) or contacting an independent third-party expert to discuss (multiple companies have mentioned that X is especially important, what is your experience in this space?) For example, I was working with a university professor who had developed an interesting software for diagnostics. Discussions with doctors resulted in a lot of pushbacks, as many did not think this technology would integrate well into existing ERPs. To get another side of the feedback, we contacted hospital IT administrators to get their perspective, and the results were quite valuable in the commercialization strategy.

On the y-axis, startups measure how compelling their product is for potential customers. Is it generating buzz? Is it challenging conventional thinking? Is it garnering strong interest? A start-up on the lower end of this axis might not have an inferior product, just one that does not really resonate with their customers.

It might be difficult for a company to measure the resonance of their product. Here, you can use the 40% rule: if at least 40% of the potential customers you talked to expressed strong interest in your product as something they absolutely need, then it is a good indicator you have good resonance. If you are lower, you might struggle to gain significant traction to reach your market. You might need to redefine your target market, or you might have to redefine your product.

The obvious question that comes next is "How do you measure strong interest?" Once again, nothing formal exists, but there are clues that you can gather. If you ask a potential client "So are you interested in my product?" and you get a positive response, it is difficult to evaluate between polite interest and real traction. Some better indications could include the client taking the lead in the interview with questions such as "This sounds really interesting, when do you expect this to come out?" or "This sounds like a cool product, are they looking for beta users? I would love to setup a trial at my organization!" In a PMF project I did, a product had such resonance that potential clients contacted me a year after the interview, inquiring on the technology progress and availability of trial products … that's strong resonance!

Another way you could measure resonance is by asking if the interviewee agrees to be contacted in the future as the product nears the market. When I do this for client, I always wrap up calls by asking "Would you like to be contacted by the company once the product nears the market to get a more detailed description of what the product does?" If I have a 50% positive response rate of interviewees agreeing to share their coordinates to the client, I know that we have good resonance.

If you are already selling your product, some practical hints that you have not achieved resonance are that word of mouth is not really spreading around your product, usage is not growing significantly, media and press interest around your product is weak, the sales cycle is too long (and it is growing longer), and many deals just never close. Inversely, if you are hiring sales and customer staff as fast as possible, sales are faster than your production, and reporters are calling to learn about your product, then you have achieved resonance.

 **Philippe weighs in on the importance of evaluating your PMF:** Tech startups tend to get overambitious with their products, especially when it comes to user experience. While there are benefits to having a complex product, such as allowing users to customize their experience, it can also lead to confusion and frustration. As such, it is important to ensure that the product is as simple and intuitive as possible.

The best way to improve the user experience or simply your product is to come face to face with reality. This is a recurring problem among "DeepTech" entrepreneurs, they seek to obtain the most successful product to address the market. Among the companies that have always impressed me are those that had succeeded with a product or service in development to set up a proof of concept with an industrialist or a key player in the market.

For example, a company that was developing a diabetes monitoring tool with both a device and an application really impressed me. One of the co-founders was himself diabetic and he knew the customer's need particularly well and he had integrated the habits of potential users into the development of the product. Also, the product was particularly integrated into the users' "use flow". It had a particularly high customer resonance with particularly interesting storytelling focusing on real assets to penetrate the market.

Coming back to our model, once you have measured both resonance and customer engagement, you can use the following framework to evaluate where you are in the PMF journey (Figure 3.1).

If you have high resonance with your market, and high customer engagement, congratulations, you have successfully achieved **PMF**! The next challenge will be maintaining yourself in this quadrant, watching out for challenges such as high-customer churn (which would suggest you are growing out of touch with your customers).

Many start-ups fall into the sexy story or revenue-centric quadrant. Having generated early sales, they figure they have found a good fit, but

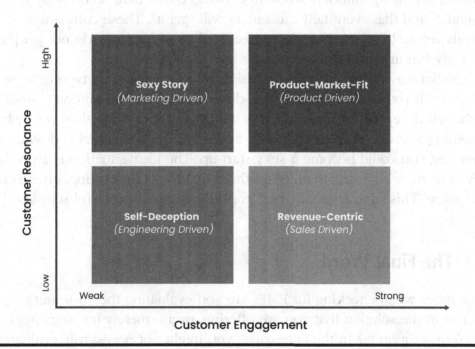

Figure 3.1   The four quadrants of product-market fit.

these customers might be outliers rather than being a manifestation of your main customer base. You might be getting the wrong feedback, or a very niche segment of the market might be excited by your product.

Companies that are **revenue-centric** are generating sales, but this is masking that they have not attained customer market fit. For example, each sale might be very customized with a high level of post-sale service, or it could be you have a very wide range of different clients. It might be fine in the short term as you bootstrap your company, but it will be difficult to scale up.

Companies that are **sexy start-ups** often do not realize they are in this position until it is too late. They might have an interesting product, that resonates well with media and investors, but find themselves unable to commercialize beyond their early adopters or are unable to monetize their early free users. While they have great feedback, they do not realize they are in an echo chamber, and that the sales that do come through have a prohibitive cost of sale per unit or are irregular. These companies substitute customers with media and investor attention, assuming customers will eventually join in.

The last quadrant is one all too familiar to experienced investors and stakeholders in the start-up space. We call these companies **self-deceiving startups**. These companies have neither customer resonance, nor customer engagement. This is typical of small early start-ups where the company is being driven by founders who are convinced that their technology is essential, and that eventually customers will "get it". These companies usually refuse to adapt to customer feedback, and normally do not get past the early funding rounds.

Another way of framing the discussion is looking at which type of team you have. If you have an engineering-driven team, you risk putting yourself in the self-deception quadrant. If sales are driving your team, then you risk becoming a revenue-centric company, and if it is marketing that is driving your company, you could become a sexy start-up. The ideal situation is a product-driven team, with assets in all other three quadrants (marketing, engineering and sales). This team is usually best placed to grow a successful start-up.

3.3 The Final Word

Remember when checking for PMF: Are you evaluating the customer's interest in the solution that you are offering, or are merely investigating a problem/solution fit? In the latter case, you might not necessarily evaluate the interest in the product or service you have developed, identifying a false positive, an interest in the market for a solution to a problem, but not necessarily that your product is the solution people are looking for.

As such, be sure to ask direct questions on your product, inquiring about some of the features that are directly relevant to your technology, not just exploring your product end goal. If possible, take advantage of your conversations to explore client sensibility to pricing, as this data will be invaluable when developing your financing models moving forward.

Notes

1. Alzate, Miriam, Arce-Urriza, Marta, and Cebollada, Javier. 2021, September 1. What are consumers saying online about your products? Mining the text of online reviews to uncover hidden features. *Journal of Digital & Social Media Marketing*, 9(2).
2. Fisher, Adam. Adam Fisher on how to navigate the product-market fit journey, 25 November, 2019. https://www.bvp.com/atlas/the-product-market-fit-journey#page_top (Last Visited 27 May 2023).

Chapter 4

Basics of Intellectual Property Strategy

Intellectual property (IP) is an essential component of start-ups in life sciences, and this chapter should not be construed as a definitive guide to this space. Rather, what my colleague and I will be sharing is a starting point so you can quickly grasp the fundamentals and be able to ask the right questions as you build your own IP strategy.

In the next few pages, we will be discussing the two main aspects of your IP strategy. First, we will present the four major components of an IP portfolio: patents, trademarks, copyrights and trade secrets. Afterward, we will present some of the activities that you can engage in to enhance the value of your IP portfolio (which is described in Section 4.2), which are the patentability assessment (4.2.1) as well as the freedom to operate search (4.2.2). We will conclude with a few thoughts on the overall value of an IP portfolio, from both an entrepreneur and an investor point of view.

4.1 The IP Portfolio

Throughout its development, a life science company will develop numerous tangible and intangible assets. These might include new inventions, a strong brand, an enviable reputation, efficient internal processes, a strong list of clients and more. These assets all have value, and if they were copied or stolen by a competitor, you could lose your competitive advantage. To protect these assets, there are legal tools that you can use to either protect

DOI: 10.4324/9781003381976-4

them or hide them from your competition. We refer to the collection of tools to protect these assets as the IP portfolio.

In the following pages, we will be discussing four main tools. Most of these exist in one form or another in most legal systems, with various levels of enforceability: as such, consulting a legal expert in the environment you are operating in is essential. Also, most of these forms of protection are protected by national agencies: as such, it will be necessary for you to repeat these actions for any jurisdiction you decide to operate and commercialize in. The four tools we will be describing here are patents, trademarks, copyrights and trade secrets. You can find a short summary of each tool in the table below (Table 4.1).

Table 4.1   Executive Summary for IP Portfolio Main Components

Component	Use	Strengths	Weakness
Patent	Gain the exclusive right to use an invention (product or process)	• Stops other parties from using or copying your invention • Makes licensing and commercialization simpler	• Information becomes public • Limited time protection • Difficult/costly to acquire and enforce
Trademark	Protect a slogan, phrase, or name	• Gain exclusive use of your brand • Can last forever if renewed on time • Makes licensing and commercialization simpler	• Protects mostly marketing and branding positioning • Difficult/costly to enforce
Copyrights	Protecting a specific work (like a song or story) from third-party use	• Gain exclusive use of your work • Long duration • Makes licensing and commercialization simpler	• Ambiguity in copyright laws: open to interpretation • Difficult/costly to enforce
Trade Secrets	Protecting information by not revealing it	• Does not cost anything in terms of filing and upkeep costs • No lifetime limitations • Protects against entities which would not respect existing patent laws	• Your innovation can be reversed engineered and copied • Weak protection in case of replication by a third party • Some jurisdictions don't recognize trade secrets

4.1.1 Patents

The first tool we will be exploring is the patent. As described by the World Intellectual Property Organization, a patent is *"an exclusive right granted for an invention, which is a product or a process that provides, in general, a new way of doing something, or offers a new technical solution to a problem. To get a patent, technical information about the invention must be disclosed to the public in a patent application"*.[1]

A patent is granted by a national body, and gives you exclusive right to use, and to exclude others to use the invention that is protected. As such, a patent can cover technical inventions such as chemical compositions (like pharmaceutical drugs), mechanical designs or new machine designs. You can also patent a new and useful improvement to an existing invention. Patents cover a specific invention, and it is not uncommon for an innovative solution to be covered by dozens of different patents simultaneously.

One particularly important nuance is that your patent does not grant you the right to use your invention, but rather the right to stop others from using it. As such, getting a patent on an untested drug does not allow you to commercialize it, and it does not let you bypass the regulatory approval process. The patent office looks at three things to grant a patent. First, that the product is novel (first of its kind). Second, that it is non-obvious (that the invention is not obvious to someone in the concerned field). Third, it must show utility (functional and operative). As such, you could demonstrate that a drug is functional and operational (it impacts the biological system of a person taking it) without knowing if it effectively cures what it is designed to target.

While you have a patent, you can stop other companies from copying, making, using or selling your invention without your consent, which is its main strength. You can also stop companies that file new patents or make new products that improve your invention without your consent. For example, if you have filed for a patent for a new chemical compound for a new pharmaceutical drug, and someone used your chemical compound to create a new different drug, you could stop them by claiming patent infringement.

Another main advantage of having a patent is that it makes the innovation easier to commercialize. Hence, it is possible to sell licenses to third parties allowing them to use your patent, and it is easier to manage this business transaction with a patent than if you only had trade secrets, for example.

There are also some disadvantages to patenting an innovation. First, to obtain your patent, you must accept to publish your innovation. This means the information becomes available to all, and competing companies might be able to copy your invention (especially if they operate in a country where patent enforcement is weak, or if you did not file for a patent specifically for this country). Second, patents have a limited life span: protection given is usually for 20 years from the moment the patent is deposited (not accepted), and while it is possible to file for some extensions to extend patent life, patent protection will eventually lapse. Finally, patents can be costly and difficult to acquire, and might require considerable resources to manage and defend.

There is a space where patent law gets especially complex: patenting software. Some argue that that is difficult to capture a software-related innovation into words: as such you might not be able to patent a software directly, but you might be able to patent a computer method that performs specific actions or steps, focusing on functionality rather than focusing on the technical aspects. Overall, if you are developing an app in artificial intelligence or a digital therapeutic solution, it might be possible to patent your innovation, but you should be conscious that it is a challenging space.

4.1.2 *Trademarks*

Trademarks are usually words (or a combination of words), sounds or designs that are used to properly identify your goods and contrast them from others in the marketplace. Some examples of elements you might be able to trademark include your company name, your company logo, the shape of your goods or its packaging, for example.

There are three main advantages to getting a trademark. First, getting a trademark will prevent others from registering the same trademark, while also helping you stop others from using your trademark (or a similar one) to sell their goods. As such, if you are developing a business model where you are spending a lot of resources to create a brand, applying for trademarks will be essential to protect your efforts and stop others from creating confusion in the marketplace.

Second, in many jurisdictions (such as the US, Europe, Japan and Canada) trademarks do not have an automatic expiry date, but you are responsible for maintaining your registration. As such (generally), every ten years, you must file certain documentation to show that you are still using your trademark. If it is approved, that trademark duration is then extended for another ten years, which means there are no legal limits to your trademark. Finally, as in patents, trademarks make commercialization (through licensing) easier as it gives a legal framework and a potential enforcement mechanism.

Nonetheless, there are some limits to trademarks. First, trademarks are often perceived as granting a limited protection, as they protect a marketing and branding position. They do not protect the goods and services themselves, but rather their names. If you have a trademark on a medical device, which a company copies and then renames to something else, you will not be able to protect yourself with a trademark. In this case, patent protection would be the way to go.

Also, as we discussed earlier for patents, enforcement and protection is up to you: if someone is infringing on your trademark, you will have to enforce your trademark and pay the litigation costs. This can be especially difficult as trademark litigation can be more subjective and subject to interpretation. Nonetheless, trademarks can be an effective component of protecting a company's brand.

4.1.3 Copyrights

Copyrights give the author the exclusive right to reuse a specific work in a public context. This might be a literary, artistic, dramatic or musical work (including those found in apps). While copyrights apply less to companies in life sciences, it is an essential element for companies developing digital healthcare solutions as well as digital therapeutics. For example, your app might have a catchy tune, or might use a series of distinctive icons to represent distinct functions. These could be covered by copyrights, to stop competitors from using your catchy tune in their own app. The length of a copyright claim can vary from one jurisdiction to another, but a good rule of thumb is 70 years from publication.

One of the main challenges with copyright enforcement is the ambiguity in the space, which can leave a lot to interpretation. For example, imagine you designed a catchy character for your digital app that shares mental health tips. A competitor then copies parts of the mannerisms and colors of the character, without copying it integrally. Your enforcement would not be as clear-cut as it would be in a patent case, for example.

4.1.4 Trade Secrets

Trade secrets are valuable information that you have decided to not make public, and instead elected to hide and keep secret. Trade secrets might include recipes, component lists, fabrication processes, as well as commercial secrets such as sales and distribution methods, competitive profiles and client lists. The Coca-Cola formula is one of the most famous illustrations of a trade secret.

Why would someone choose to protect their innovation by using a trade secret? For some, protecting their innovation through a trade secret is a temporary measure, and is used while the company is applying for a patent. Other companies have limited funds and must choose which innovation to patent and which one to protect with trade secrets. Some companies fear that they might be copied quite easily or believe that the value of the innovation might exceed the 20 years offered by patent protection. Finally, some innovations might not meet the patentability criteria, forcing the company to rely on trade secrets instead.

If you choose to use trade secrets, there are a few things you can do to secure your secrets. First, use non-disclosure and confidentiality agreements whenever you disclose information. Second, include confidentiality clauses in your employee agreements. Finally, use encryption, password protection and locks to lock up your valuable information.

To qualify as a trade secret, information must 1) have commercial value, 2) be kept secret and 3) reasonable efforts must be made to keep the information secret. Being able to demonstrate these three points is essential if you engage in litigation.

Even if trade secrets are free, and can be implemented simply, there are several disadvantages to using trade secrets instead of applying for a patent. First, if your innovation is reverse-engineered (dissected and analyzed by a competing company), they might discover your technology secret and be able to use it. Also, your trade secret, if developed independently, may be patented by another company first, significantly hindering your ability to license (and even use) your trade secret. Finally, once the trade secret is made public, anyone may have access to it, and use it accordingly.

This brings up another interesting point, which is the overall protection of trade secrets. The main protection aspect of a trade secret being secrecy, there is little protection if someone develops the innovation independently. Trade secret protection can only be enforced if you can demonstrate illicit acquisition, or use/disclosure of confidential information. Even then, enforcement of trade secrets is weak in some jurisdictions, especially when compared to the protection conferred by a patent.

Hence, remember that once a trade secret is known to the public, it is no longer protected. While you might believe you are well protected, there are several situations that could occur leading you to reveal your trade secrets. These include writing a scientific publication that reveals just a bit too much, disclosure of information during a presentation with parties not covered by the non-disclosure agreement, conversations during a trade show, accidental disclosure through a misdirected email and so on.

 **Philippe weighs in on value and perception of IP for investors:** An IP right is an asset that provides its owner with a form of exclusivity and commercial exclusivity. Such exclusivity may be conferred by law, as with patents, trademarks, or copyrights, or may be arranged by the owner, based on secrecy, as with trade secrets and know-how. It is an asset by the very reason of this exclusivity which provides the owner with a competitive advantage.

To start, there are a few things to keep in mind. First, intangible assets are the least liquid assets of a company. Among these intangible assets, IP rights, especially patents, are frequently considered the most liquid. Currently, there is no exchange or highly transactional market like real estate for IP rights.

A patent confers on its owner an exclusive right only if it is granted and if it is maintained in an appropriate manner by paying the annuities. A patent application does not confer an exclusive right. Finally, for reference, a 2008 study showed that only 20% of European patents granted could claim a value greater than or equal to €3 million, the sum of the value of these 20% representing 90% of the total value of all European patents issued and active on that date. However, most of the time, an assessment relates to a portfolio of IP rights and not a single patent. Also, for a life science investor, the patent is the most robust IP tool. It is a major asset. Finally, in the due diligence of potential acquirers of technology companies, the due diligence of IP is particularly stringent.

4.2 Activities That Enhance the Value of Your IP Portfolio

There are several activities that a company can engage in that can add value to your IP portfolio. We will be going over two of them that can generate value in an investment scenario, as well establishing the legitimacy of your portfolio: a patentability assessment, as well as a freedom to operate search.

4.2.1 Patentability Assessment

A patentability assessment is made prior to engaging in the full patenting process. It is done to assess the likelihood of obtaining a patent for an invention and getting an idea of what it could cover. As such, before investing fully into a patent, a company can get a preliminary idea of the potential of its patent. It is more than just a patent search (which is research of prior art done to understand the scope of available patent space). A patent assessment goes into more details, and while it includes a patent search, it can also include an opinion on the other requirements of a patent such as novelty and utility.

A patent assessment will usually give an opinion on the patentability of an invention. This can be useful in both product development, as well as in situations where multiple patent opportunities exist, and the founders must select where to invest their limited funds. While more expensive than a patent search, it is also more valuable in terms of the information that it makes available to concerned parties. It is a strategic document that can be useful before making a major patenting investment, as well as enhancing the overall value of your patent.

4.2.2 Freedom to Operate Search

Patenting your innovation is only the first step to protect your innovation. Once you start to develop an IP portfolio, it becomes important to ascertain what you are allowed to do with these patents. As a reminder, a patent gives you a right to exclude others from making the claimed invention, it does not confer authorization to make, use or sell the invention. As such, you do not have the guarantee you can exploit your innovation without limitation. That is where the freedom to operate (FTO) search comes in.

An FTO details your company's ability to develop, make and market your invention using your patent without the risk or liabilities. It assesses the patent landscape around your own IP and identifies any IP that could keep you from using your own IP freely. Without this document, you might be violating someone's IP without even knowing it. Life sciences is a patent-focused space, and it is often crowded with a lot of overlapping patents and prior art. As such, an in-depth FTO is essential to moving forward.

FTO analysis usually occurs prior to an investment to ensure the real value of the patents being presented. In these cases, it is common for investors to request an FTO prior to investing. It is also done prior to selling assets to ensure their overall use and demonstrate their overall value. Companies can also do a "tree analysis" FTO to assist in technology and development decisions, targeting the space with the least third-party IP inhibiting development and avoiding litigation or licensing royalties: this can be done early in the development phase so you can plan and modify product development around potential patent barriers that are uncovered.

Philippe weighs in on the value of in-house IP evaluations: Some start-ups I have evaluated in the past decided to do this type of work in-house. From a VC perspective, if you are doing it for your own research, that is fine. It also shows some initiative, which is a nice touch. **Nonetheless, from a due diligence and investment evaluation perspective, there is only value in an FTO if it was done by a third party.**

An FTO search can also include a patent map, which is a graphical method of visualizing patents in the relevant space. It helps decision-makers visualize and understand the patents, players, and interactions between them, and can be a useful tool to illustrate the current situation. To help you understand, we have built a simulated patent map (figure 4.1) for a hypothetical company, called Vaxxon.

The larger rings represent the most relevant entities to this specific map, while the lines represent agreements, patents or challenges between two entities. Smaller rings represent entities that are identified as being relevant to the underlying therapeutic technology. By a quick look, we can identify main players, as well as licenses and litigation issues.

In our example, Vaxxon Vaccines is a small biotech company developing vaccines in the polio space. It is currently engaged in patent litigation

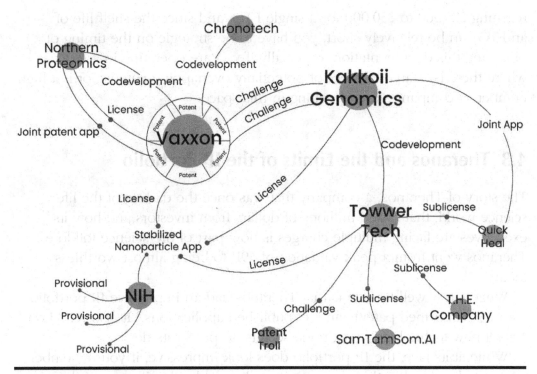

Figure 4.1   Simulation of a patent map.

with Kakkoii Genomics, while co-developing technology with Northern Proteomics and Chronotech. The NIH, which has developed a stabilized nanoparticles delivery application, has licensed the technology to Vaxxon, Kakkoii and Towwer Tech. Kakkoii Genomics is also licensing technology, as well as collaborating with Towwer Tech, which could lead to potential conflicts.

An FTO search is also useful in case you are accused of patent infringement: as life sciences is a space where litigation over patents is quite frequent, having an FTO detailing a perceived operating space will be an essential component of your defense strategy. A carefully done FTO search can be invaluable in defending against accusations as it demonstrates your company's efforts in respecting other people's IP. In the US, a finding of "wilful infringement" can result in the damage awarded being tripled.[2]

Finally, it's important to underline that an FTO search is not a guarantee that you are not infringing on any other patents, but rather a legal opinion based on the analysis of the existing information that is currently available. As such, not all companies get an FTO for each of their patents. With costs reaching $25,000 to $50,000 for a single FTO, and since the shelf life of an FTO can be relatively short, you have to be strategic on the timing of obtaining this documentation, especially if you are operating in a space where there is a large number of potentially overlapping patents, or if a high number of companies are operating in this space.

4.3 Theranos and the Limits of the IP Portfolio

The story of Theranos, a company that was once the darling of the life science world, that raised millions of dollars from investors and how its executives are facing multiple charges is now part of life science folklore. Theranos went from a peak valuation of $9B to being almost worthless today.

What is less well known is that Theranos had an impressive IP portfolio, with 137 confirmed patents and 47 published applications.[3] It also relied on "secret new technology"[4] (i.e., trade secrets) to power its device.

While at its face, the IP portfolio does look impressive, if you remember what we discussed in this chapter, you will quickly identify two of the potential gaps in the Theranos story. First, **patents do not demonstrate that the technology is functional**, but rather that there is a reasonable assumption that it does. Many investors were burned based on the assumption that having all these patents validated the technology. Second, as the nature of the trade secret is to remain confidential, **there is little corroboration or validation of a trade secret's validity**. As such, trade secrets usually are of limited value to investors when they review an investment opportunity, and being able to demonstrate the value and reality of your trade secret will be key in having the investor accepting it as a component adding value to your company.

Philippe weighs in on the value of trade secrets from an investor's perspective: When doing a first evaluation of a start-up, I usually give little value to trade secrets. They are hard to evaluate, and as such, difficult to give it a monetary value. I prefer demonstrable know-how.

As such, the lesson is your IP portfolio is a necessary part of your story to get the conversation started with investors but be ready to defend your innovation based on results. IP is now being scrutinized a lot more thoroughly than before, and companies no longer get a free pass for "having a lot of IP".

4.4 Final Thoughts

 From the perspective of the entrepreneur, a well-constructed IP portfolio is an invaluable asset when building your company. Obtaining patents enables the company to show the value of your innovation and sets you up for future licensing opportunities. But the overall strength of your IP portfolio will not always be based on the quality of the documentation, but rather on your ability to enforce them. If you have limited resources to monitor, litigate and enforce your IP, your patents might be encroached, and you will have little you can do. **Monitoring and enforcement of your patents is up to you, as there is no "IP Police" monitoring patent infringement.** Furthermore, failure to enforce patents might erode their overall value: If a party can show that you have failed to enforce your patent in the past, it might limit the overall damage you can collect. Finally, some regions of the world have very weak patent enforcement mechanisms, so for some technology, a mix of IP and trade secrets might be worthwhile but validating mechanisms (such as the patent assessment and FTO) are quickly becoming investment staples during due diligence.

 Also, if your technology is originating from a research laboratory, it is key that the license of the technology be compatible on the one hand with the gold standards of the industry and with terms (upfront, royalties, milestones) consistent with market practices. During multiple due diligences, I have been embarrassed to find that licenses with TTOs (Technology Transfer Offices) have been badly negotiated. It is essential to approach an IP consultant to support you in the negotiation to obtain an acceptable license corresponding to the market standard.

Notes

1. World Intellectual Property Organization. What is a patent? https://www.wipo.int/patents/en/ (Last visited 15 April 2023).
2. Morgan Lewis, Chapter 7 – *Freedom to Operate, Emerging Life Sciences Companies*, Second Edition, page 57.
3. Stratford Group – Could a better understanding of patents have saved investors from Theranos? https://web.archive.org/web/20220626053312/https://stratford.group/could-a-better-undertsanding-of-patents-have-saved-investors-from-investing-in-theranos/ (Last visited 27 May 2023).
4. Diamandis, Eleftherios P, Lackner, Karl J, and Plebani, Mario. 2022. "Theranos revisited: the trial and lessons learned." *Clinical Chemistry and Laboratory Medicine (CCLM)*, 60(1), 4–6. https://doi.org/10.1515/cclm-2021-0994 (Last visited 27 May 2023).

Chapter 5

Developing a Customer-Centric Approach

 Some measure the success of a technology innovation by evaluating how it has improved an existing situation. For example, we examine if the innovation has enabled us to process information faster or more efficiently than existing technologies, or if it can detect a condition which other technologies are not able to. But when we focus uniquely on technology development, we omit market insight. As such, it is common to attend an interesting technology presentation, and start the feedback session by asking the founders:

- "**Who** will use this innovation? **Do they need** it? Are they **looking for it**?"
- "**Who** will pay for this new innovation?"
- "Is it **significantly better** than existing products to justify changing what they already use/already have?"
- "How will this technology **integrate** into existing processes?"

For example, a few years ago, I worked with a client who had developed a technology that analyzed a laboratory sample significantly faster than existing technologies. In a vacuum, this seemed like a great selling point for end-users, and could be sold as a great competitive advantage. After doing interviews with potential users, I found that the current technology (which took several hours to analyze the sample) was not a bottleneck in the lab: during the process, technicians would let the old technology "do its thing" by itself while they moved on to other tasks. The innovative technology, while faster, held

DOI: 10.4324/9781003381976-5

little interest to them as it did not address any pain points. Yes, laboratory managers might purchase the device if they did not have any device capable of accomplishing this task, they had little interest in recommending it for purchase to replace existing lab equipment. Current workflow processes had developed around the current technology, and there was no interest in disrupting it since it created little value. As such, a technology that was demonstrably faster, hence obviously better than existing technologies, drew little market interest from its potential customers. A new positioning around higher accuracy and precision had to be developed.

This is why we believe that customer-centric product design is crucial in creating value. Customer-centric product design occurs when product development incorporates client feedback directly into its development process in a continuous manner. This ensures that the development team focuses on designing a product that will not only improve the current work situation, but it can also successfully be integrated into existing processes while complying with current user workflow.

As we will explore in the next few pages, in life sciences, the customer who is paying for the innovation can often be a different party than the end-user using the technology, so you might be developing a product that, while solving pain points for the end-users, is of no additional value to the party paying for the product. As such, successful companies must correctly map the customer process to understand who uses the product, who influences purchasing and who does the actual purchase, while also understanding the overlap between the three functions. They must also be able to create value for all interested parties in the value chain, not only the payer or the end-user.

This chapter will address three topics: first, we will present an overview of the healthcare ecosystem, followed by an illustration of a few payers' systems and a clear description of key stakeholders. These are necessary to understand the concept of clients and can help map the purchasing process. We will close the section by discussing customer-centric design and by illustrating the parallels between the traditional product development and a more customer-centric product development.

5.1 Introduction to the Healthcare Ecosystem

5.1.1 Overview of the Healthcare Ecosystem

The traditional healthcare ecosystem usually comprises four main components.[1] At its the core, we have the individual patient, who is taken care of by his care team. This team includes professional care providers

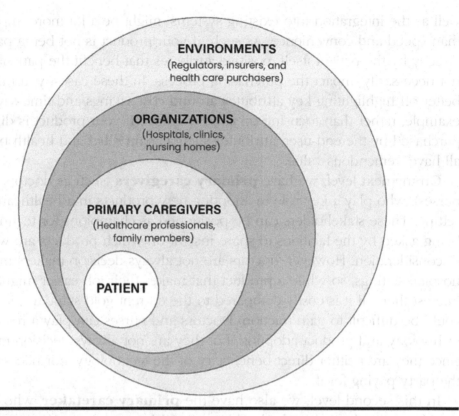

ENVIRONMENTS
(Regulators, insurers, and
health care purchasers)

ORGANIZATIONS
(Hospitals, clinics,
nursing homes)

PRIMARY CAREGIVERS
(Healthcare professionals,
family members)

PATIENT

Figure 5.1   Illustration of the traditional healthcare system.

(such as clinicians, pharmacists and other healthcare professionals) as well as family members. Encompassing them, we have the organization that supports care (e.g., the hospital, clinics and nursing homes), which itself is impacted by a political and economic environment (e.g., regulatory framework, financial system, payment regimes and market dynamics for example). These define the conditions under which organizations, care teams, individual patients and individual care providers operate. This is illustrated in Figure 5.1.

In our first bubble, we have **the patient**, who is (hopefully) at the center of the healthcare system. In many cases, he is the end-user of a product and, as such, is often perceived by innovators as a client. Hence, product development and marketing strategies are built with the patient in mind. For example, devices that perform diagnostics quicker are often believed to be invaluable for the hospital setting since the patient will be diagnosed earlier and have an answer faster to his current situation.

Unfortunately, in this situation, patients' preferences might have negligible impact on which product the healthcare provider will purchase. Pricing, as well as the integration into existing systems, might be a lot more important than speed and convenience. As such, if your product is not being paid directly by the patient itself, positive attributes that benefit the patient do not necessarily impact the purchasing process. In these cases, you might be better off highlighting key attributes around cost savings and time-saving, for example, rather than focusing on client comfort. If your product is directly purchased by the end-user, attributes such as pain relief and health benefits all have tremendous value.

On the next level, we have **primary caregivers** (such as doctors and nurses), who play a key role in adopting new products in a healthcare setting. These stakeholders can be positioned as champions for technology, being asked by the facilities to share insight on which products are worthy of consideration. However, doctors are not always decision-makers in hospital settings, so while a product that makes their job easier might interest them, if it is costly compared to the current gold standard, it could be difficult to gain traction. Doctors and nurses can play a role in technology and product adoption, but they are not always decision-makers since they are neither direct beneficiary of the technology, nor necessarily the party paying for it.

In this second level, we also have the **primary caretaker**, who, while not always implicated in the healthcare decision, can be an important influencer in the purchasing decision. A caregiver is usually defined as someone from the patient's network that helps them with daily activities. The objective is to help preserve the person's autonomy and ensure that they can recover. Caregivers are often involved in situations of long-term care, disability care or old age custody. They will usually be sensible to services and products that can assist them in their tasks, which might not immediately benefit the patient. Products that help track patient movements around a house or reduce burden when doing recurrent tasks (ex: monitoring heart rates) are simple examples of products that could target caregivers.

In our third level, we have the infrastructure that supports healthcare. **Hospitals** are organizations dedicated to giving care to patients in need. They can be for-profit or non-profit. Depending on their relationship with GPOs (group purchasing organizations) and insurance companies, they can have various levels of latitude when selecting products. Either way, they will often have purchasing processes and departments dedicated to the procurement of products, selecting products and services based on request for quotations.

Also included in the infrastructure are clinics and nursing homes. **Private clinics** usually have more flexibility around purchasing innovative technologies and are often targeted as first clients for innovations, as they are often better equipped for small-scale integration and can be leveraged as successful success stories. As for **nursing homes**, they usually have limited budgets for innovation, but will be receptive to technologies with potential for cutting costs or improving safety. They are also great environments to test relevant technologies on a limited basis if a comprehensive monitoring strategy and support plan is proposed during presentation.

On the last level, we have the group environment around innovation, which includes regulators, insurers, and GPOs.

Regulators play a significant role in defining the rules around delivery of healthcare (what can and cannot be done), and often interact with companies when evaluating the safety of the devices being developed. Ensuring that your products obtain all the certifications is an enormous responsibility and is an important cost center depending on how extensive your approval process is, and how regulated your device or product is.

Private and public insurers play a significant role in healthcare as they reimburse part or the totality of a treatment, product or device they have specifically selected to cover. As such, products which they refuse to cover often have tremendous difficulty integrating into healthcare markets. The role of insurers will vary a lot from one market to another, but from a general perspective, it is important to understand that insurance companies have their own evaluation processes to select which services and products they will reimburse and will usually favor those that reduce costs. The products that are covered vary from one insurer to the next, and biological processes (which are especially expensive) are not always covered. Finally, remember that insurance companies will only cover the cost of a product or service they have initially approved, so if an alternative product is used, they might only cover the cost of the equivalent product, or even nothing at all. You can find more information on insurance systems in Section 5.2.

Group purchasing organizations (GPOs) are organizations combining several like-minded healthcare organizations, leveraging a combined group purchasing power to lower the price of a product or service due to higher expected volumes. The objective is to generate savings for its members, aggregating large volumes. One of the crucial issues when dealing with GPOs is that their involvement can add some level of rigidity to purchasing procedures. In some GPOs, memberships are constraining agreements: if the GPO negotiates an agreement to purchase a specific product, all members will be constrained to purchase the same product. GPOs will often play a role in selecting products for its members, and as such it will be important to get a read of the environment and see if your type of product is subject to purchasing restrictions.

The framework we just reviewed has several advantages. First, it puts the patient at the center of the intervention. It is also simple to understand and enables a quick and easy understanding of how healthcare works. It also highlights a basic model of how each actor is impacted by those above it. Nonetheless, it does have some limiting factors, as it underplays the interaction between stakeholders. Also, in our specific sector of interest, it downplays the role of innovation, and the role of innovation agents around the patient.

5.1.2 A Transforming Healthcare Ecosystem

We have seen an accelerating amount of innovation in healthcare settings. This was in part accelerated by COVID-19 but was also done to address mounting costs in our healthcare system. As the aging population increases, traditional ways of delivering healthcare are getting more expensive (in terms of cost and labor). Overall, three main trends have led to the transformation of the healthcare system.

First, innovation is being driven by advances in digital technology, which are completely transforming the healthcare ecosystem. This has resulted in companies and organizations increasingly investing in digital transformations to exploit data across channels, operations and patient outreach.[2] How innovative companies manage and secure the data they generate data will be an important criterion for investors, regulators and purchasers when evaluating technologies.

One of the more popular applications of digital technologies has been telemedicine, which is considered a solution to many challenges currently impacting healthcare; thanks to telemedicine, it is possible to bridge geographical distances, and deliver care to both underserved populations while accelerating access for individuals. Challenges going forward in this space include developing digital technologies that help doctors in diagnostics, which is currently limited. As one doctor I interviewed in the space once said *"I can talk with the patient, ask questions, but whenever it gets a bit more complicated, I cannot really do much but do a referral. I can help with smaller issues, but anything complex sends the patient back to the in-person clinic"*. IT infrastructures also play a key role in the adoption of telemedicine. It is often quite a paradox, as regions who would benefit the most from telemedicine (being remote and underserved from a healthcare perspective), and are the ones where the IT infrastructure is less developed, slowing down the implementation of these technologies.

Furthermore, because of lifestyle changes and emerging technologies, we are witnessing a push toward preventive and proactive healthcare, as well as a growing trend toward self-care. As such, we are observing the emergence of a person-centered healthcare system focused on wellness and not disease, with an increasingly aware and self-reliant patient, driven and accompanied by a series of innovations in technology and business models that will change healthcare as we are used to understanding it.[3]

This is true, not only for economically advanced nations countries but also for low-income countries as they will face the same issues requiring more hospital care, but with limited infrastructure which will result in some treatment pathways moving out of hospitals.[4]

This evolution has also led the patient to seek care less often in traditional settings and the consequent expansion of alternative care sites. Patients are now able to receive care where and when they prefer through a variety of flexible formats. With various authorities moving forward with unique patient records, it will soon be possible for each facility to be informed about the patient's medical record regardless of where the patient has requested treatment.[5]

As innovation plays a key role in transforming the healthcare system, it has led to the acceleration of mechanisms and organizations to accelerate the development and integration of these technologies. Two key trends have been important in this area. First, the growth in the number of accelerator and incubator programs around the world. Multiple established stakeholders, from universities to local governments to private companies have been implementing programs that are essential to accompany entrepreneurs in developing their technologies. Simultaneously, hospitals have been deploying efforts to create dedicated personnel and offices to help guide and integrate technologies, matching entrepreneurs with key personnel. As such, these trends are both seen as a positive sign that technology integration can be handled in a professional and efficient manner.

 **Philippe weighs in on the evolution of the healthcare system:** Driven by the health crisis, the digital transformation of "e-health" is accelerating. Startups participate fully with their innovative technologies: from telemedicine software to solutions boosted with artificial intelligence and connected objects. Overall, they are creating an array of innovations that appeal to investors.

According to a study prepared by the Institut Montaigne[6] (Associated with the firm McKinsey), the deployment of e-health could generate between 16 and 22 billion euros per year while designing health system of tomorrow. Mobile applications, e-prescriptions, online appointment booking are all technologies with a lot of potential. While part of the French territory is being reconfigured, digital solutions in the service of health are proving their usefulness. But the sector still needs to accelerate its digital shift.

The COVID-19 crisis demonstrated quite aptly the low digitization of the French healthcare system. It highlighted several successes and dysfunctions of the French health system, especially when compared with other countries which had vastly different responses to this epidemic. Nonetheless, according to a recent analysis by venture capital players. In 2020, 84 European funds invested in e-health (26 in France), including 14 dedicated funds (six in France). The most active countries in the sector are France, followed by the United Kingdom and Germany.[7] So, France has many funds investing in e-health. A growing sector, thanks to the COVID-19 crisis. Enough to boost startups in the sector in the development of their projects.

This evolution has led to a more distributed healthcare ecosystem, with more stakeholders having a hand in innovation. From charities and associations funding innovation directly (through the emergence of charity-backed VCs) to insurers investing in innovation and acting as testing ground for innovative technologies, start-ups must now contend with a much more eclectic environment (Figure 5.2).

Overall, while the patient stays at the center of the healthcare system, the number of stakeholders interested and impacting innovation has increased.

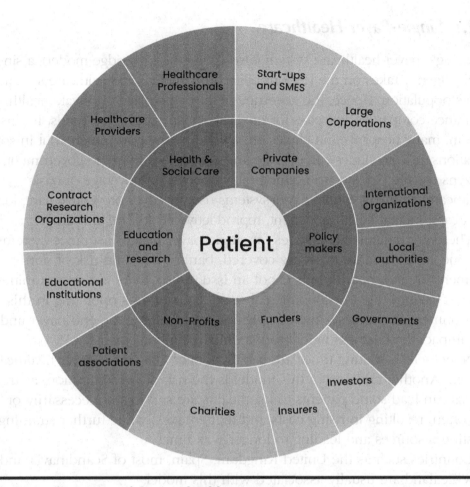

Figure 5.2   Illustration of the new healthcare system.

5.2 Different Ecosystems, Different Payer Models

On our journey to understand the client, it is important to understand how the insurance system is structured and understand how payments are made in your target market. While we will be sharing an overview of the four basic models (single-payer, social insurance, national healthcare and out-of-pocket), it is important to emphasize that national healthcare systems are usually hybrids, and careful research of your target market will be necessary to understand the environment you will be operating in.

5.2.1 Single-Payer Healthcare

In a single-payer healthcare system (also called the Beveridge model), a single public agency takes on the responsibility of financing the healthcare system for its population. Hence, the government takes the place of private health insurance companies and pays for medical activities using tax funds. In this system, many hospitals and clinics are owned by the government, but in some situations, private clinics can operate and collect fees from the government. Citizens usually pay little to nothing out of pocket for coverage of basic healthcare treatment. Single-payer systems usually cover preventive care, long-term care, mental health treatment, reproductive healthcare and drugs.

There are multiple advantages for single-payer health insurance systems. For one, as all medical costs are covered, bankruptcy (and risk of non-payment) due to medical bills is not an issue. Also, there are opportunities for cost savings due to economies of scale. For start-ups operating in this environment, it decreases the complexity of identifying the end payer and can impact the business model to maximize reimbursements.

Nonetheless, waiting times are usually longer in a single-payer healthcare system. Another criticism of this model is the risk of overutilization, as free access can lead some patients to use healthcare services unnecessarily or too often, resulting in rising costs and higher taxes, while further straining existing resources and leading to longer wait times.

Countries such as the United Kingdom, Spain, most of Scandinavia and New Zealand are usually associated with this model.

5.2.2 The Social Insurance Model

Also called the Bismarck model, the social insurance model is a decentralized form of healthcare where the responsibility of healthcare payments is shifted to employers and employees. In this system, the two parties become co-responsible for funding health insurance companies through sickness funds funded by payroll deductions. Private insurance plans cover every employed person, regardless of pre-existing conditions, and the plans do not make a profit.

Providers and hospitals are private entities while insurers are public companies. In some countries, there is a single insurer (France). In other countries, like Germany and the Czech Republic, there are multiple competing insurers. In all the examples above, the government controls pricing.

Countries using these models usually get faster and better access to care, as well as more consumer-oriented care. The primary criticism of this model is that care is only afforded to those who can provide funds. Those who are unable to work or cannot afford contributions (such as aging populations) get limited coverage. As the cost of insurance continues to grow, more of the population becomes underinsured.

The social insurance model is used by Germany, Belgium, Japan, Switzerland, the Netherlands, Latin America and to some extent, France.

In fact, the French model (like many models when examined more carefully) is complex, holding on to some elements of both the single-payer healthcare and the social insurance model. It is financed by a government national health insurance, which is paid by the population, which deducts a premium directly from the employees' paychecks, rather than being funded through income tax. It is a co-pay system, where the social security system will cover 70% of the global cost the patient paying the rest of the charge. As such, while it is categorized as a social insurance model, it includes many of the characteristics akin to the other models examined.

Philippe weighs in on the evolution of the healthcare system in France: As Jean-François mentioned, the French healthcare system is quite complex. The Social Security system (in French, système de sécurité sociale) is divided into five branches: illness, old age/retirement, family, work accident and occupational disease. The French social security is managed by many organizations. In 2019, highlights for health expenditure related to the consumption of medical care and goods (CSBM) was estimated at 208 billion euros.[8] So broadly speaking, Social Security financed 78% of the CSBM, and complementary organizations (mutual funds, insurance companies and provident institutions) contributed about 14%. France devoted a total of 11.3% of its national wealth to health, i.e., 1.4 points more than the European Union average. It is the OECD country where households are the least financially challenged!

Axa, a French multinational insurance company (100 billion euros revenue in 2021) is really looking for solutions to provide to its customer. I was discussing with Thomas Buberl, the CEO of Axa and clearly the possibility of offering digital monitoring solution for Axa customers with chronic diseases such as diabetes is key. One model under discussion with e-health key French payers was the opportunity to provide devices and digital solutions to insured customers. The insurer would provide for free digital solution devices to monitor insured/customer and according to the behavior of the patient, the insurer will offer discounted cost for health cover. To give an idea of the challenge of being able to monitor diabetic patients, it is estimated that 7.7 billion euros (8,3 billion USD) are spent in France each year on complications related to diabetes and which could often be avoided thanks to better data monitoring patients' daily lives.

5.2.3 National Health Insurance Model

A national health insurance (NHI) model includes a mix of public and private healthcare providers, being paid by a single government-run insurance program that every citizen pays into (through their taxes). In theory, a mix of both private and public providers operate in the healthcare environment, giving more flexibility for patients without having to navigate complex insurance plans and policies.

Universal coverage is the main advantage of a national health insurance model: attempts are made to cover all citizens, regardless of salary and revenues. Another advantage is having a single-payer gives the NHI system considerable market power to negotiate lower prices. Nonetheless, critics to NHI models include the potential for long waiting periods as well as potentially encouraging waste and even fraud.

Canada, Taiwan and South Korea are the main examples of this type of healthcare model.

Canada's NHI reflects both the strengths and weaknesses of this model. The federal government, while responsible for funding the healthcare program, delegates the application of healthcare services to provinces. As such, all ten provinces operate independently, with limited interaction with the other ones. Many services are funded through federal funds, but provinces also contribute, and they may charge a health premium for non-essential services. Nationally, this has led to debates on what constitutes essential and non-essential services. As for private care providers, they operate in periphery of the main healthcare system, offering services which are not always covered or reimbursed by the public insurance.

5.2.4 The Out-of-Pocket Model

The number of countries with a structured healthcare system is in the minority and might account for less than 20% of the world's countries. While many countries are trying to implement centralized models, most countries rely on the out-of-pocket model for the delivery of healthcare.

In this model, patients must pay for healthcare services out of pocket (directly to the healthcare provider). As such, the wealthiest population gets access to professional care, while the poorer population must get by using substandard care. Out-of-pocket care occurs in rural regions of India, China, Africa, South America, where patients might go their whole life without ever seeing a doctor.

5.2.5 Standing on Its Own: The United States

When we present the four basic models, they look monolithic and easy to understand when observed individually. Unfortunately, many countries will be hybrid systems, incorporating two (or more) systems into its healthcare system. The United States is a great system to illustrate a system having multiple frameworks co-existing in a single healthcare system. The US maintains a complex relationship with its healthcare system and is the only industrialized nation that does not provide universal healthcare coverage.

Single-payer healthcare is available in the US for veterans (i.e., retired soldiers): Care to veterans is provided by the Veterans Health Administration (VHA), America's largest integrated healthcare system, which provides care at 1,298 healthcare facilities, including 171 medical centers and 1,113 outpatient sites of care of varying complexity (VHA outpatient clinics).[9] Through this extensive healthcare network, the VHA provides inpatient and outpatient care at VHA medical facilities, prescription drugs from VHA providers, long-term care and mental health care without cost to qualified candidates.

The social insurance model is available for Americans who get insurance as part of their job benefits. Employer-sponsored health insurance is paid for by businesses for their employees as part of their employee benefit package. Nearly all large employers in America offer group health insurance to their employees.

As for the national health insurance model, it is called the Medicare system. Medicare is a health insurance program provided to people 65 or older, which will pay the bills of its participants using covered facilities. While patients pay part of their costs, it does ensure some form of funding for more vulnerable populations.

Finally, 10 to 12% of the US population have no health insurance and must pay their bills out-of-pocket at the time of treatment. As such, uninsured Americans usually choose to postpone medical treatment to avoid getting into heavy debt.

5.3 Correctly Identifying Your Client

Having taken a moment to define the healthcare ecosystem, its main stakeholders as well as the payment providers, we can now address the main topic of this section: Identifying your client. Fundamentally, your customer is the individual or organization that will be **purchasing** your product or service. He is also called many other things such as the consumer, the buyer, the purchaser or the client, but these different words describe the same phenomena.

There is often a disconnect in life sciences between the purchaser (who pays for the product), the end-user (who uses the product) and the decision maker (who decides which product will be purchased). Sometimes, the same individual fills all three roles, while in other situations, up to three different individuals interact in the same transaction (Figure 5.3).

To illustrate the complexity, let us start by a quite simple non-life science example. Let us say you are selling a pair of gloves. You advertise said gloves on television. The consumer, after viewing your commercial, decides to visit your website to purchase them, or visit a local store to purchase them directly. In this situation, the end-user, the decision-maker and the payer are all the same.

Now, let us make the process a bit more complex. Say you are selling surgical gloves like those used in operating rooms. The end-users are

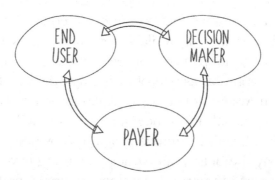

Figure 5.3 **Relationships in purchasing healthcare products.**

surgeons and nurses in the operating room. So, do you advertise to them? It depends. On one side, it is highly unlikely that they are going to be the payers for this type of product. So, while they might not purchase them directly, they might influence the internal purchasing processes. But already, there is a disconnection between the user and the payer, while the decision-making responsibility could be shared.

Now let us make our example a bit more complex. Let us say you have designed a diagnostic machine that completes a diagnostic in 5 minutes but costs $100 of consumables each time it is used. Current hospital costs are $10 for each diagnostic but take 12 hours to complete. For the consumer (the patient) the convenience is immeasurable, but since he is not the person purchasing the product and he is not necessarily the payer, his needs and opinion are negligible compared to the savings being generated to the hospital through the traditional method.

As such, there are four main types of client relationships in life sciences (Figure 5.3):

a. The relationship between the end-user and the decision-maker: A patient (end-user) needs a medication for his condition. His doctor (decision-maker) tells him to use brand ABC rather than using another brand. The patient listens to the advice and decides to purchase it.
b. The relationship between end user and the payer: A patient (end-user) needs a tooth implant. The insurance company (payer) only reimburses bridges. The client chooses to use this procedure instead of a full implant.
c. The relationship between the payer and the decision maker: A doctor (decision-maker) recommends changing from one medical product to another. The hospital (payer) acquiesces, recognizing the doctor's expertise on the topic, and purchase the new product instead.
d. The relationship between the end-user, the decision maker and the payer: A doctor (end-user) prefers a brand of medical glove. The hospital (payer), which pays for the product, while sympathetic to the doctor's preference, is constrained in his contract with his GPO (decision-maker) to purchase another brand.

There are two takeaways from this brief overview:

a. Identify who is paying for your products or service
b. Map the purchasing process for your product

Table 5.1 Who Is Your Client?

Key Question	End-User	Decision-Maker	Payer
Who is paying for the product?			
Who is using for the product?			
Who is deciding who purchases the product?			

You must understand if the purchasing is done independently of the end user, if it includes recommendations from third parties or if end users' opinions are sought out. This will have a direct impact on how you develop your product, and how you address your end users. For example, if clients are an insurance company more concerned about integration into existing infrastructure, you might need to develop close ties and relationships with them during development, whereas a company which prioritize uptake in usage and participation might be more concerned with end-user feedback and design. Understanding this purchasing cycle will be key in developing a customer-centric culture. Identifying your client can be done by answering the three questions found in Table 5.1.

The objective of Table 5.1 is to quickly identify the commonalities between the different possible contributors to the process and identify how many different entities are involved in the process. Answering those three questions will often be extremely useful to define your commercial model.

5.3.1 Building Customer Profiles

Customer profiles are fictional personas that you build that represent typical customers that purchase your products. Building these profiles helps you better understand them, what makes your product unique for them, and is useful when preparing your marketing strategy, your advertising, choosing your distribution channel and so on. Each time you are preparing your tactics, you have a sounding board against which to check your assumptions.

For example, if you just prepared an ad, you can take a second look at it through your customer's eye, asking questions such as:

- Would my customer like this ad?
- Will it make him look for more information?
- Will it be a call to action?

Building personas will require an in-depth understanding of your clients. This will mean performing secondary research, supplemented by one-to-one interviews, and site observations if that is possible. You want to get as acquainted as possible with your shortlisted best potential clients.

Once your information is collected, you can start building the persona. Some key information you typically compile and develop includes:

- Typical title/role in the organization
- Goals and motivation
- Challenges and pain points
- Where do they learn about new products and services?
- What is the best way to engage with them?
- What is their purchasing process?
- Do they have purchasing authority?
- Purchasing customer behavior?

For example, let's go through a theoretical persona. In our scenario, the company has developed an innovative technology that is used in labs and is looking to better understand its potential customers.

Using the profile found on the next page, we can see some of the current paint points are around sensitivity and quality control. Also, engagement is best performed by on-site visits, and demos, rather than doing web-based contacts or traditional advertising. Finally, purchases are done around cost, and while Terry the Tech Lab is not the decision-maker, she is an important influencer (Figure 5.4).

5.4 Developing a Customer-Centric Approach

Once the client you should target is identified, you can correctly plan and build your development strategy. One of the first things you should do is include your client into your product development. Involving consumers in the design process at an early stage and in an ongoing fashion seems logical to do because clients are, in fact, the people who will buy the product. Having the product users involved ensures the product is easy to use and fits its purpose, so that clients will not only use it, but will also enhance the likelihood they encourage others to purchase it.

The traditional way of developing products, the product-centric approach, is more vulnerable than ever. Companies used to focus on design,

Terry – The Typical LabTech

About
- Lab Technician in a University Research Center
- Has been in the same position for the last 4 years
- Performs experiences in the lab every week and handles day-to-day tasks
- Uses technologies that are currently available in the lab without questioning them

Goal and motivation
- Wants to successfully completes her experiments
- Needs to insure results are accurate and reproducible
- Showcase and promote her research in high profile publications

Challenges and pain points
- Technologies currently in the lab have sensitivity issues
- Has challenges in quality control

Where does she learn about new products and services?
- Typically relies on vendors for information on new technology
- Does attend 2-3 conferences per year, but does not focus on new tech
- Reads generic science magazines (Nature, PubMed), and those relevant to her specific field of research.

What is the best way to engage with her?
- Likes it when vendors come to her lab with the "new toys" so she can try them out
- Once a month is the regular patterns she expects
- She quickly discards emails, pays more attention to phone call and personal visits
- Advertising has little to no impact

What is the purchasing process in her organization?
- New purchases must be approved by lab director
- While she does not have purchasing authority, she usually makes recommendation for purchasing
- Purchases follow the economic model – Cost is very important when choosing products.

Figure 5.4   Example of a customer persona: Terry – The typical lab tech.

manufacturing and logistics. In the past when products and services could achieve a clear product/service difference, sustainable and beneficial, a product-centric approach made sense (Figure 5.5).

Hence, in traditional model, patients are usually involved in development of health products or services in the early and final stages of a project. For example, a company might interview clients in the early process to guide development. Once the product has reached a stable version, they might reach out to potential customers via usability testing

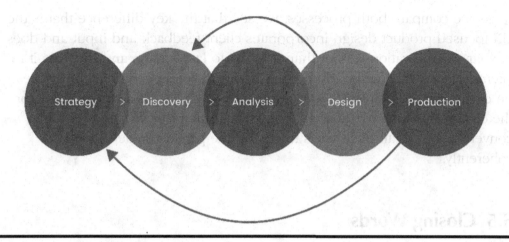

Figure 5.5   Traditional product design.

to evaluate their product or service. However, involving patients during development is essential to improve a product before it becomes too costly to do so.

Furthermore, making a customer-centric product means that not only customer acquisition will be easier, but so will customer retention.

Hence, there is a distinct advantage in co-production of products with clients, rather than relying on their input when it might be too late in their production cycle to incorporate suggestions and critical changes. Agile companies understand that the people who use their products and services have assets which can help to improve those services, rather than simply needs which must be met. Remember that patients (and often their caretakers) have "lived the disease" and intimately know the day-to-day implications and pain points (Figure 5.6).[10]

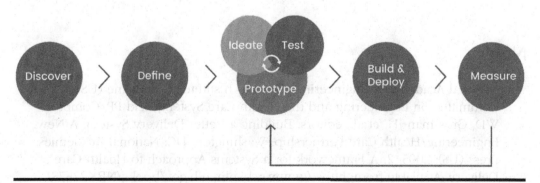

Figure 5.6   UX focused product design.

As we compare both processes, we see that the key difference that is the UX-focused product design incorporates client feedback and input and does regular small iterations rather than arriving to last version and showing it to clients. Hence, during the testing phase, the customer-centric team focuses on designing and delivering a user experience that is intuitive and easy for the customer to use. Methods like usability testing, surveys and customer conversations are all great ways to ensure the product is developed coherently.

5.5 Closing Words

When you use a customer-centric approach to innovation, you solve problems for the user while you build, rather than having long development steps. You can catch mistakes or missed opportunities quickly and can adjust your product without overinvesting.

Including potential clients in your product development (rather than waiting to get to the end of your development cycles) is not only beneficial to your development. One can make an argument that developing your product up to certain milestones can accelerate commercialization and enhance fundraising. It also limits various issues such as "feature creep" (when more features are added to a product in a mindset to respond to customer feedback) or "design limbo" when a product constantly goes back to the design table, unable to get to a state where development is sufficient to display the technology.

The work around here is to pre-plan your development with target milestones and set targets for moving forward. Sometimes you need to publish so you can stop fiddling with the punctuation, and you must move forward to build and sell your product.

Notes

1. National Academy of Engineering (US) and Institute of Medicine (US) Committee on Engineering and the Health Care System; Reid PP, Compton WD, Grossman JH, et al., editors. Building a Better Delivery System: A New Engineering/Health Care Partnership. Washington, DC: National Academies Press (US); 2005, 2, A Framework for a Systems Approach to Health Care Delivery. Available from: https://www.ncbi.nlm.nih.gov/books/NBK22878/ (Last visited 10 April 2023).

2. Russo Spena, T, and Cristina, M. (2020), "Practising innovation in the healthcare ecosystem: the agency of third-party actors." *Journal of Business & Industrial Marketing*, *35*(3), 390–403. https://doi.org/10.1108/JBIM-01-2019-0048 (Last visited 10 April 2023).
3. Aceto, G, Persico, V, and Pescape, A. (2018). The role of information and communication technologies in healthcare: Taxonomies, perspectives, and challenges. *Journal of Network and Computer Applications*, *107*, 125–154. https://doi.org/10.1016/j.jnca.2018.02.008 (Last visited: 10 April 2023).
4. Schiavone, F, and Ferretti, M. 2021 Dec. The Future of healthcare. *Futures*, *134*, 102849. doi: 10.1016/j.futures.2021.102849. Epub 24 September 2021. PMID: 34584276; PMCID: PMC8461037. (Last visited 1 April 2023).
5. Schiavone, F, and Ferretti, M. 2021. The future of healthcare. *Futures*, *134*, 102849. doi: 10.1016/j.futures.2021.102849. Epub 24 September 2021. (Last visited: 10 April 2023).
6. Institut Montaigne. E-Santé: Augmentons la dose! Published June 2020. https://www.institutmontaigne.org/publications/e-sante-augmentons-la-dose (Last visited 10 April 2023).
7. Hacquin, Arnaud. Karista cartographie les fonds européens d'investissement actifs dans la santé numérique, April 2021. Website: https://www.esante.tech/karista-cartographie-les-fonds-europeens-dinvestissement-actifs-dans-la-sante-numerique/ (Last visited: 11 May 2023).
8. Panoramas de la Drees, Les dépenses de santé en 2019 – Résultats des comptes de la santé – Édition 2020, 15th of September 2020, https://drees.solidarites-sante.gouv.fr/publications-documents-de-reference/panoramas-de-la-drees/les-depenses-de-sante-en-2019-resultats (Last visited 10 April 2023]
9. The US Department of Veterans Affairs, https://www.va.gov/ (Last visited 10 April 2023).
10. Gilbert, RM. (2022) Reimagining digital healthcare with a patient-centric approach: The role of user experience (UX) research. *Frontiers in Digital Health*, *4*, 899976, https://doi.org/10.3389/fdgth.2022.899976 (Last visited 10 April 2023).

Chapter 6

Developing Your Business Model

 Have you ever listened to a company presentation at a conference, and walked away thinking *"What does this company do?"* or *"How do they expect to make money?"* Both Philippe and I have. Maybe the presentation was too focused on the science. Maybe the management team was unable to articulate its commercial strategy. Or the company had not put any thought into how it would commercialize its technology. In the end, the results are the same, as a good opportunity gets weighted down by an incomplete business model.

ARTICULATING YOUR BUSINESS MODEL IS ESSENTIAL TO DEMONSTRATE YOU ARE READY FOR BUSINESS.

So, what is a business model? It is how you plan to structure your commercial operations so you can generate revenues from your innovation. We articulate this model by making three key decisions:

1. Defining your commercialization model: How will you interact with your end-users? How are you going to structure your relationship with them? Are you going to be in direct contact with them, or through a third party? This is the topic of Section 6.1 – The commercialization model.

 DOI: 10.4324/9781003381976-6

2. Defining your revenue strategy: How will you generate revenue? How do you plan to monetize the relationships with your client? We will explore these topics in Section 6.2 – The revenue model.

3. Defining your corporate strategy: What are you trying to accomplish? What will your key activities be? Which partners will be working with you? We will explore these topics in Section 6.3 – The corporate strategy.

I will close this section with a simple example of a typical innovation, a reflection on the different models it could adopt, and propose a pathway to commercialization.

6.1 The Commercialization Model

As we saw in the previous chapter, customer-centric product design is key in developing both your product and your company. As such, we spent a lot of time describing how we can define the client. In the next few pages, we will be spending more time on how we interface with both our end-users and clients, and the different models on how to monetize this relationship. In this section, we will be bridging from defining your client (Chapter 5) to defining your relationship with them (Figure 6.1).

In the next few pages, we will look at the classic models of business-to-business (B2B) and business-to-consumer (B2C) (Section 6.1.1) as well as business-to-government (B2G) (Section 6.1.2), and then showcase their

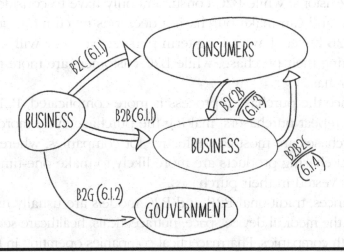

Figure 6.1   Traditional and innovative business models.

evolutionary models such as business-to-consumer-to-business (B2C2B) (Section 6.1.3) and business-to-business-to-consumers (B2B2C) (Section 6.1.4).

6.1.1 Classic Models: The B2B and B2C Models

Companies that choose a **business-to-business (B2B)** model focus on providing services or products to other businesses, whereas companies that use a **business-to-consumer (B2C)** model sell directly to individual consumers. Companies that choose a B2B model generate a lower volume of transactions but can sell at an overall higher price. As such, this model is used when you have a product where you can focus more on long-term goals and objectives, or when the product purchasing decision is more complex. In doing so, you must understand that you are engaging in longer sales cycles.

In the B2B model, you will be building personal relationships with potential clients, and focus on demonstrating your product to potential purchasers. As such, identifying the right person to interact with will be critical. As for the B2C model, trying to develop many customers means that you will have a lot less personalization and less direct relationships with your client, and you will focus more on mass-market communication tools. The greatest advantage here is that having a diversified customer base can translate in faster growth, as most individual purchase decisions are less complex.

B2B commerce is more complex than the B2C model, as the purchasing decision is longer, much more intricate and involves multiple individuals. For example, B2B buyers must consult with multiple departments before purchasing, get budget approvals and are held responsible for their purchasing decisions, while B2C consumers only have to consider their own interests, and can make purchasing decisions much more quickly. Also, since B2B buyers have a long-term perspective, they will spend more time researching their purchase, while B2C customers are more prone to impulsive purchases.

Finally, since the purchasing process is more complicated, B2B purchasers are generally repeat purchasers, making the purchase cycle more profitable as repeat purchases are most cost-effective for companies, whereas individuals purchasing products are more likely to make one-time purchases, and have less vested in their purchases.

In life sciences, traditional B2B and B2C models are usually used by companies in the medical device space, nutraceuticals, healthcare services as well as digital health companies. Pharmaceutical companies operating in the over-the-counter/non-prescription space are also likely to engage in a B2C model.

6.1.2 Eschewing the Private Sector: The B2G Model

The **business to government (B2G)** is a model where a company selects (or is compelled by existing regulations) to sell and market goods and services directly to government agencies. The model is often more complex than the traditional B2B and B2C models, as it is very bureaucratic, and requires strict compliance to many laws and regulations due to additional oversight and monitoring.

In many situations, functioning in this sector means responding to pre-negotiated contracts, request for tenders or other bidding mechanisms. While selling to B2B is more complex than B2C, B2G is often a lot longer than both as it goes through multiple approval and vetting processes. They also generate a lot of additional paperwork and reporting requirements, and payment cycles are typically longer than those of private companies. Nonetheless, these programs are often larger and more stable than private sector contracts and are often less risky than contracts with the private sector (as government agencies seldom go bankrupt). Furthermore, once a company is vetted by a government agency, getting future contracts can go faster (as the company has already gone through the vetting process). Finally, some government agencies will sometimes have "My First Customer" type programs, where a company in its first steps of commercialization is favored in tenders and given preferential access to a market to stimulate early growth.

B2G in healthcare often means working on contracts with public hospitals and healthcare systems. Pharmaceuticals, biotechnology and medical devices are those more likely to engage in B2G commercialization.

6.1.3 Evolution in Digital Health: The B2C2B Model

As models evolve, we have seen a push in life sciences (specifically in digital health), towards the **business-to-consumer-to-business (B2C2B)** model. In this model, a company starts by targeting individual consumers. In turn, these consumers are encouraged to promote the product or service to their employer from the inside, with the end goal being to convert the company to a corporate customer. The app, which starts being commercialized using a Freemium Model (see Section 6.2.4), evolves to monthly payments based on corporate key metrics such as enrollment, engagement or even health outcomes such as number of steps, or weight loss.

This model is gaining favor in the digital health sector as it hinges on two key trends: on one side, consumers are increasingly invested in their own healthcare spending, making them a good target for rapid adoption campaigns. As for companies, they are getting new apps pitched to them on an increasing frequency, which has created some fatigue in evaluating new opportunities. Hence, apps that are "pitched' from the inside are increasingly favored. This model also has the additional benefit for the innovating company of creating valuable usage data, as app providers can show current successful ROI of employees using the app, which they can then leverage in their business development activities.

The greatest advantage of this approach is that it enables shorter sales cycles than B2B through demonstrated user traction and user results, while taking advantage of the quick traction it is possible to create through a B2C campaign. From an investor perspective, this model has the advantage of demonstrating multiple customer traction metrics (adoption, retention, velocity) while also demonstrating early enterprise validation (showing a pipeline of buyers, case studies and even signed contracts).

6.1.4 Using Companies to Reach Individuals: The B2B2C Model

Some entrepreneurs approach the market from another angle, first targeting companies with the end goal of reaching consumers. This model is called the **business-to-business-to-consumers (B2B2C)**. In this model, companies want to leverage the recognition gained from having a larger company adopt a product, as a bridge to build trust with consumers, leveraging the larger consumer base of the B2B client. In this model, a profitable partnership could lead to many customers with an extremely low per-customer acquisition cost, as going directly to B2C can be very costly to a small company.

Medical devices and digital health are the most likely to use this model. For example, imagine having designed an app to monitor a medical condition. The company could have a relationship with a private insurer, which accepts to pay the monthly cost for the first year. After a year of use, if the consumer has seen enough value from the technology, they could decide to pursue the business relationship and continue paying the fee, creating a direct relationship with the medical software company.

Some good practices in this space include building good relationships with your business clients, and being able to demonstrate that while your goal might be to develop a relationship with your client's clients, your goal is not to take them away. For example, developing a business relationship with an insurance company, and highlighting your model as being company agnostic, enabling clients to move from one insurance company to the next easily might not create excitement in potential clients.

6.1.5 *One Product, Multiple Models*

To better illustrate our different models, let's imagine a hypothetical company, DigiHeart, deciding on its commercialization model.

In our example, DigiHeart has designed a health app that helps patients with cardiac conditions, capturing health data and sending it to the Electronic Health Record (EHR), making it available to the patient, his doctor and his care team. Also, the app, through a series of webinars, gamification tools and reminders has shown an overall benefit in improving patient health. DigiHeart now must commercialize the product, and the end-user is clearly patients with a cardiac condition. But how to approach these end users?

In the B2B model, the target is other organizations. In fact, large corporations have shown a strong appetite with partnering with small agile digital health developers, especially in the fields of chronic condition management (such as digital therapeutics, sensors and wearables, and remote patient monitoring), care navigation (helping provide patient advocacy, member engagement, utilization management, and care coordination) and digital benefits (such as self-service tools, paperless communications, process automation and clinical/EMR integration). As such, DigiHeart could decide to partner with insurance companies, who will then make the app available to the people they insure, or they could target an employer, as a way to help its employees optimize care. In both cases, the emphasis is on the reduction of medical costs while providing superior care, all the while emphasizing long-term relationships with the client company.

A second option might be to forego businesses and targeting the consumers, using a B2C model. As such, the DigiHeart could decide to engage in a social media advertising campaign, making the APP available in app stores, with the goal of quickly ramping up patient use. This strategy focuses on increased patient self-health trends, as well as the increased use of medical devices for self-monitoring.

In a third option, DigiHeart might decide to use a B2C2B model, ramping up individual users rapidly, then leveraging them through the release of new features locked behind corporate sponsorships. For example, giving clients the ability to create a group of individuals from the same company to share tips or compete against each other, might encourage them to contact their company to sign up in a corporate account. As such, the APP might have several individual features, and lock collaborative features behind a paywall, to encourage users to get their company "on board".

In another model, DigiHeart might decide instead to develop corporate relationships with larger companies, with the overall goal of building individual relationships with the employees (B2B2C). In this model, the companies could sign on for a basic package available to all employees, with individual employees free to unlock specific modules as they see fit. They could also family packages, so employees could offer significant rebates to friends and family.

In the last model, DigiHeart could target a government entity (B2G). In this model, DigiHeart could convince the government agency to bear all costs related to the APP, while the APP is distributed freely to all relevant users. In this case, a state-level entity could decide to make the app available to its citizens as part of a global campaign on reducing healthcare risks relative to cardiac conditions.

All these options have merit and have distinct advantages in terms of speed to market, generating revenues, margins, and so on. In any case, your presentation will have to emphasize on two things. First, why was this model chosen for commercialization? Once this is answered, you will have to address some of the steps the company will have to take to put this commercial model into operation.

6.2 Revenue Model

In this section, we will be going over the most popular revenue models used by companies in life sciences. Some of these may not be applicable to your company: for example, the sale of products in the pharmaceutical and biotechnology space is heavily regulated, and is often subject to existing price structures, and as such, companies that sell these products must respect existing price codes, while not directly advertising to end-users.

Also, in some cases, you may find that using only one model is not appropriate, and you might need at adjust, or even combine two to achieve profitability. You could also plan to have a two-tiered system, where you use a model for growth, and a second one for longer term.

As you prepare your presentation, it will be essential to document why you choose these strategies, the expected revenue for each model as well as well as which milestones would trigger the shift from one model to another.

6.2.1 The Transactional Model

The transaction model is the most traditional revenue model, whereas a product or service is sold at a fixed price and purchased by the client (but not necessarily the end-user). It is by far the most direct form of revenue model, and one for which a company has the most control. The main advantage is that this model can then be customized to include various strategies such as bundling, quantity discounts and price differentiation. The main disadvantage is that engaging in direct transactions implicates higher involvement and more investments, from elaborating a pricing strategy to delivering customer service.

As an example, imagine a nutraceutical company that sells Vitamin supplements. These supplements, sold through their website, are sold at a fixed price. The company is now responsible for driving customers to their website, and as such must work on an advertising strategy, as well as a customer service department. Nonetheless, as they are in control of the selling activity, the company could decide to give a discount if a customer purchases two distinct types of supplements (bundling) if a client purchases multiple items of the same product (quantity discounts) or giving him a rebate if he is a returning customer (price differentiation).

Finally, as mentioned earlier, you could select a transactional model where you do not directly interact with the customer, rather using a third-party intermediary such as a distributor or a company with a large sales force. In this case, the relationship is still transactional as you are selling your products directly to a company, and all the other considerations (such as pricing strategy and customer relationship management) come into play.

6.2.2 Membership/Subscription Fee

A subscription revenue model is one where you offer a product or service to a customer, which can be paid on a monthly, quarterly, biannually or even annually depending on how you want to structure it.

You could also add pricing tiers, where some users get access to more product offerings if they are paying a premium. When used to sell a software, you can refer to this model as a SaaS model, or Subscription as a Service model.

The difference with this model (versus the transactional model) is that the relationship would not emphasize the purchase of the product, but rather the nature of the relationship you have with your client. As such, your product could be part of the bundled purchase, rather than the objective of the purchasing relationship. Also, companies that employ a membership model often focus on a long-term purchasing relationship, rather than targeting one-time purchases.

The advantage of this model is that the company is developing a steady revenue stream. Revenues emerging from transactional models can be aleatory, while building a revenue stream that focuses on memberships can generate recurring revenues, enabling you to better plan your ongoing costs and expansions. Nonetheless, this model is dependent on having a large consumer base, and it is critical to maintain high subscription and renewal rates to ensure steady revenue flows. This means putting higher emphasis on newer features, newer content and upgraded devices, something that a company having one-time purchases has less pressure to do.

Digital platforms are the main users of this type of revenue model, but it is also used by some medical devices (mostly those with consumable components). It is also used by companies in the nutraceutical space. For example, imagine designing a meditation application. You could charge a basic price per month, giving access to basic content. You could also have a premium subscription offer, where users have access to extended content, and offer rebates to users signing up for a longer period.

6.2.3 Pay-per-Use Model

Some companies deploy a pay-per-use model, where the product itself is given away for free to a customer, yet requires the customers to pay for installation, customization, training and/or consumables. As such, customers generate revenues on a per-user basis. The photocopier industry is a notable example of this model, where copier can be rented to sites, and pricing is based on direct usage, as calculated by the photocopier itself.

In life sciences, this model is mostly used by medical devices (where the main device is cheap, and consumables must be paid up) and digital health technologies.

The advantage of this model is that it helps ramps up usage by clients who are hesitant to first try a product and can also be used by the client to estimate usage, while also increasing brand awareness. The disadvantage is that low usage could severely impact profitability, and that you are running a services business with the product as a marketing cost.

Digital health technologies are big proponents of this model. For example, some companies let the users install their products for free on a system but will then charge a set amount each time the APP is used to do an analysis or to generate a report on a patient. Medical devices are also a potential user of this model, especially those with consumable components.

6.2.4 Freemium Model

The freemium model is used to emphasize quick ramp-up on the number of users, and to increase product visibility. In this model, a company's basic services are offered for free, but users must pay for additional premium features, extensions, functions and so on. For instance, if you are building a service-based business or are active in the SaaS (software as a service) space, then the freemium model might be worth considering, since it lets your users try your product before they buy.

The main advantage of the freemium model is that it offers the opportunity for the client to try the product or service for free, giving them the opportunity to try out the product or service while simultaneously enticing them to pay for it later. The disadvantage is that this model requires a considerable investment of time and money to reach out to a significant part of the audience, and even more effort to convert free users into paying customers.

6.2.5 Licensing Model

For some companies, selling a product is not the objective, and commercialization is done through the licensing agreement. For these companies, consumers and end-users take on a vastly distinct perspective, as they are unlikely to ever sell a final product. Instead, they focus on monetizing their technology by selling licenses to other companies, giving them the authorization to integrate their technology into an existing product.

This model is especially favored in situations where you have developed an enabling technology (which enhances a process) or a very specialized innovation. For example, imagine, you have developed an app which, with a picture, could diagnose with certainty contact dermatitis. It is unlikely an end-user, with a skin disease, would download an app for contact dermatitis, and then another for psoriasis, a third for rosacea...rather than developing a specialized app for contact dermatitis, a much better model might be to licensing your diagnostic technology to generalist apps than operate in the skincare space.

Payment for license deals can be up-front payments, ongoing royalty payments or a combination of both. Up-front payment can be an advantage depending on your financial situation, but in my experience, you should always strive to get a combination of both, as deals structured with up-front payments uniquely don't allow the app developer to profit from product success, while royalty-only structure put the risk of generating revenues back on your end: if your partner does not successfully sell its own product, your own revenues are directly impacted. Furthermore, you could potentially be constrained by a non-performing license, and unable to sell your license to another party. Exclusivity should only be given if the financial upside is significant, or if the partner is already successfully commercializing its own product, ensuring some level of revenues mid to long term.

6.3 Corporate Strategy

The third component to decide is the corporate strategy of the organization you will develop. Hence, let us take a moment to examine some of the more popular business models in life sciences such as the fully integrated model, the virtual model, the platform model and the No Research Development Only model. Figure 6.2 (found on page 86) gives a good overview of where the different models fit in the development process.

6.3.1 Fully Integrated Model

The classical model in life sciences is the **fully integrated model,** where the company attempts to integrate all business functions under a single corporate umbrella, from discovery and development to manufacturing to sales and marketing. This model is also referred to as a "bench-to-market company". In biotech, some companies generate early revenues by licensing out a few compounds, and then selecting a few other products to commercialize themselves. Others will temporarily engage in contract-research, to maximize existing infrastructure and resources. The fully integrated model was once a more popular model for start-ups, but with time, investors and management teams have found that there are several issues with this model. First, it is quite costly to own and operate all the equipment your company might need, to fund all the clinical trials, and requires many types of distinct expertise (from researchers, to production, to managers to sales staff). Furthermore, companies that operate using this model carry heavy fixed costs. While it is less likely today for therapeutic start-up companies to aim for a fully integrated model, some of the big original biotech companies (such as Gilead, Amgen or Genentech) are from this mold. The model is not really viable for small emerging biotech companies, but it is possible for companies in the nutraceutical and medical device space to utilize this model due to a shorter clinical approval process and "simpler" commercialization roadmaps. Nonetheless, the elevated fixed costs, the risk of commercialization and dilution of expertise concerns remain.

6.3.2 *The Virtual Model*

A popular model right now in life science companies is the **virtual model**. These are companies which, quite early, decide which segment of the value chain they have expertise in, and then develop the company to fill that specific function. Usually, these companies focus on the discovery and development functions, but there are virtual companies that can focus on marketing and commercialization, for example. As such, a virtual life science company could decide to focus on discovery and development, out-licensing products as these get regulatory approval, and starting research again on new products. Having a virtual model limits costs and risk, as you usually have less fixed cost, reducing your burn rate. Nonetheless, this model is not only upside, as outsourcing some of your activities means that time and effort will have to be spent managing these relationships, and differences in culture and communication issues could disrupt product development, production and commercialization.

A typical virtual life science company will have one or two full-time employees and will then complete the team with several consultants who specialize each in their space of expertise. In recent years, we have seen the rise of "fractional CFOs" and "fractional CSO", individual who lend their expertise to a start-up on a part-time and contractual basis. As such, it is not rare to meet an individual who is a fractional CxO for multiple companies simultaneously. As the nature of the work pertains a lot to temporary work relationships (people come and go as their expertise is required), communications and project management become essential to a successful virtual life science company.

6.3.3 The Research Model

The **research model** is a variant of the virtual model. Rather than bringing the product through all the clinical phases, these companies focus exclusively on the research and preclinical phases, licensing out compounds to bigger companies that can then focus on the clinical phase. As such, these companies will usually have a strong science component in their infrastructure, with a few individuals (most likely the CEO) focusing on licensing the technology to other third parties. This model allows the company to contain costs even further as they seldom engage in clinical research, but the potential royalties of licenses are much lower as products are usually out-licensed quite early in their development phase. This is the type of company that often comes out of academia, is funded by government grants and often has only a single product in its pipeline. It is also usually typically led by a professor or a graduate student who has led the innovation, is intimately familiar with it and has a focused product development vision.

6.3.4 The Platform Model

Platform companies are another variation of the research model company. There are three common characteristics to platform companies: they have access to a differentiated technology to discover new therapeutics, have strong protection around their technologies, their technologies are scalable and their targets are agnostics.[1] These companies have three main income streams: drug revenue income (where they develop and commercialize their own pipelines) partnership income (from co-development deals) and service fee income (where they let third-party companies use their platform to optimize their own pipelines).

During the first phase of its existence, the company focuses on partnering and commercializing its platform with companies to generate revenues, shifting to its own pipeline once it has sufficient capital and traction.

Moderna is a great recent example of a platform company, initially spinning off several asset-centric companies including Valera, Caperna, Onkaido and Elpidera, and then transforming into the "drug developer" company we are familiar with today.[2]

6.3.5 *The No Research Development Only Model*

Another life science company model is the **No Research Development Only (NRDO)** model, whereas a company focuses exclusively on the clinical development of products. It licenses in products that have cleared the preclinical phase and focus on bringing them to the end of the phase III, where they will license them out to big pharmaceutical companies. They usually generate revenues through royalties and licenses.

Jazz Pharma (www.jazzpharma.com) is a notable example of a NRDO company. Active in the neuroscience and oncology space, they focus exclusively in partnering products that have completed preclinical studies, and, as per their website "have a team of experienced leaders with a demonstrated history of successful deals and partnerships, and we have significant capital available for deployment".[3]

6.3.6 *Other Specialized Models*

Other more specialized models include the **recovery model** (focusing on products that failed advanced clinical studies due to weak efficacy, and re-purposing them for new indications), the **combination model** (combining two drugs for new indications, which are then sold to big pharma) and the **drug delivery company** (which focuses on developing novel delivery mechanism for existing drugs).

In diagnostics, companies use various models as well:

1. They can decide to develop their diagnostic and license their technology for manufacturing and commercialization, akin to the research model;
2. They can decide to license their test to an established diagnostic company, so it can be paired and included in an existing device;
3. The company can decide to go to market with their own diagnostic device and commercializing it (akin to the fully integrated model)

6.3.7 *A Few Words on How to Select Your Model*

Overall, selecting your model will be determined by how far in the commercial process you want to go. For companies interested in focusing on the research space, the research model might be the one for you. For those interested in keeping all aspects of your process in-house, the fully integrated model might be for you. As always, how you justify your model will be key with the investors and partners.

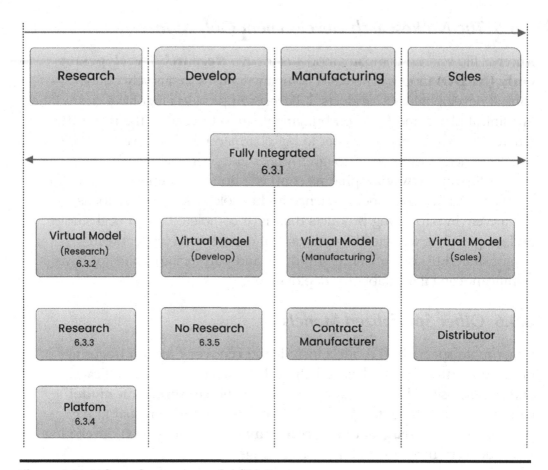

Figure 6.2 Where does your model fit in?

6.4 Planning VCs Exit

An exit strategy is an essential part of your business plan. From the very start of your business venture, you should know how you, your team and your investors plan to move on. Why? Because exits are when founders and investors get paid! It is key to plan your exit from the first day of the start-up. If you want to accelerate your company quickly, if you have substantial investments to make in your technology, if you are a deep-tech startup, you are going to have to raise funds. Or you already have. And investors, whatever they tell you, are necessarily interested in the exit scenario because that's how they are measured in term of performance (it's the famous phrase "give me performance"). Internal rate of return (IRR) for VCs is the expected annualized return a fund will generate based on a series of cash flows over the duration of the fund, which is typically ten years. IRR is used by Limited Partners (LPs) in Venture Capital to benchmark a fund's performance against relevant peer groups. LPs will consider the time since the initial cash outflow of a fund and compare it against the timeline of similar funds in the same asset class.

The main exit routes are quite limited in practice: the IPO (Initial Public Offering), mergers and acquisitions (M&A), the purchase of shares from investors at a later stage (share buyout), and to a lesser extent, strategic partnerships and royalty-based agreements. Table 6.1 (found below) shares the basics of each exit route.

Even if the IPO remains an often-invoked exit strategy, nine of ten exits are M&A. So, a common exit strategy for life science investors is a merger or

Table 6.1 Executive Summary for Main Exit Strategies

Exit Strategy	Advantages	Disadvantages
IPO	• High return on investment if successful	• IPO market is cyclical in nature. • Substantial capital investment requirements
M&A	• Faster exit timeline. • Return to investors can be substantial with reduced timeline and dilution compared to an IPO.	• Dependent on acquirer's interest and strategy
Share buyout from later stage investors	• Timelier exit	• Dependent on the dynamics of the financing market. • Implies that the valuation is higher than the cost of acquisition of the shares
Partnerships	• Provides funding to the company. • Longer term exit strategy if the partnership leads to the acquisition of the company	• Does not necessarily lead to an exit. • May even reduce other industry acquisition opportunities
Royalty-based license agreements	• Long-term source of income if no other exit opportunities available	• Long-term exit strategy and dependent on successful product licensing and sales

acquisition. This is almost always the primary exit strategy, so companies are often built to be sold, where companies communicate early and regularly with potential strategic acquirers and position themselves to be sold in the future. This provides the opportunity for a faster exit timeline than an IPO. The return to investors can be substantial with reduced timeline and dilution compared to an IPO.

The first tip is to identify potential buyers and identify previous transactions that they were involved in. This crucial step helps to understand the M&A behavior of potential buyers. The objective here is also to get an idea of the exit valuation and the critical size to be reached in order to hope to generate interest from this type of buyer.

This information will then allow you to build an intelligent medium- to long-term strategy based on attractive valuations for a potential buyer. We must not let ourselves be overwhelmed by excessively high valuations which will be a brake for the future. It is also essential to identify potential competitors on the target market and consider a partnership approach in the case of complementary solutions.

Finally, it is important to be able to create value at the key stages of the project. Your priority must be to become the benchmark in a highly targeted vertical in your market. Thus, one of the very important missions for the founders throughout the life of the startup will be to communicate. It is necessary to maintain the relationship with other corporations, in particular with potential buyers identified earlier: give regular news on the progress of the startup, highlight targeted KPIs and tell its story by creating the benchmark and assuming industry leadership.

Philippe weighs in of communication to potential partner: Too often, founders wait to have a perfect product to communicate or to set up proof of concepts with manufacturers or industrials. Even if you must be vigilant in terms of intellectual property, it is key to connect with potential buyers to create opportunities to appreciate your technology. Without having feedback from internal technical teams, industrial M&A teams cannot position themselves on an acquisition operation. In fact, even if it seems counterintuitive, it is important that the co-founders can put themselves in the following posture: "the company is never for sale, but it is always for sale".

6.5 Bringing It All Together, an Example of a Hypothetical Company

To properly illustrate, let us take a classic example of a company spinning out of a university, going from an academic project to a fully independent company.

In this example, we have two young Ph.D. students who have developed a medical device that they believe will revolutionize the diabetic monitoring industry with an application and a tracking device for diabetic patients. It does patient monitoring and shares advices and solutions with an application including nutritional data. Through a web dashboard that translates key blood glucose, carb and medication data into statistics and graphs, providers are better able to manage their patients with diabetes.

They have completed their first prototype and have negotiated with the university tech transfer office, so they have access to all relevant IP that was generated, against the payment of a reasonable license fee. As they build their company, they have three decisions to make:

a. What will our commercialization model be?
b. What will our revenue model be?
c. What corporate strategy are we going to emphasize?

From a commercialization standpoint, they select a B2B2C model. As such, having both worked part-time in the insurance industry, they believe they have the necessary connections to be able to connect the insurance industry, being able to talk their language and build a compelling case for these companies (hence the B2B component). They also believe that once insurance company clients benefit from the technology, end-users will be well-positioned to demonstrate the technology to relatives and friends, hence the B2C component.

As for the revenue model, the founders decided to keep the model simple and focus on a transactional model. The founders have obtained some quotations for manufacturing, and there are significant discounts for ordering larger quantities. As such, quantity discounts become a key component of their revenue model. Furthermore, a product-market fit has demonstrated both interest and acceptability for the current price range.

Finally, the founders enjoy designing device products, and already have several ideas in the pipeline for future innovations. They have little interest in selling the products themselves, as well as manufacturing or distributing them. As such, they decided to design a virtual company with a small R&D force, outsourcing both the sale and manufacturing of their products. In the end, they prepare the following three slides, which they present to an established VC (Figure 6.3).

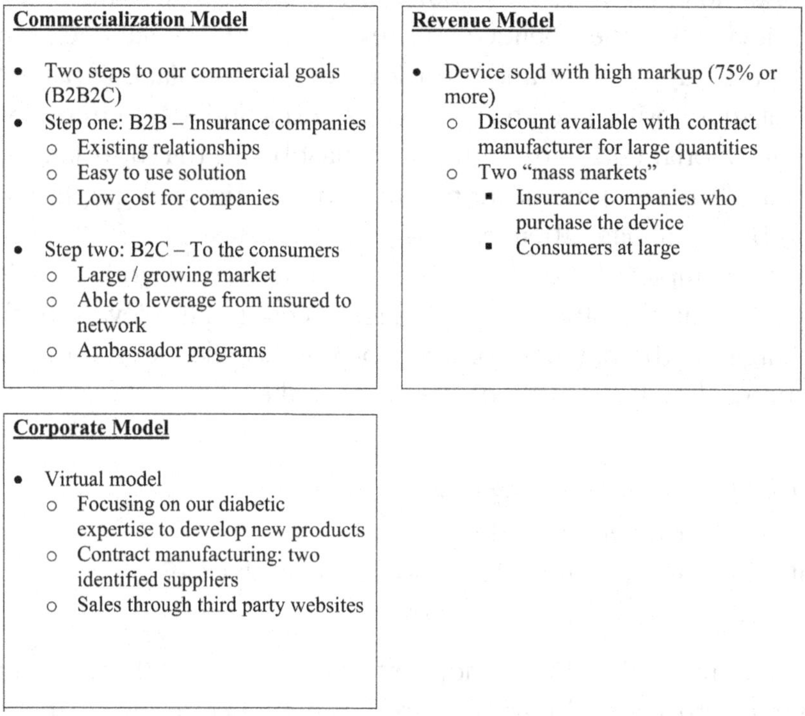

Commercialization Model

- Two steps to our commercial goals (B2B2C)
- Step one: B2B – Insurance companies
  - o Existing relationships
  - o Easy to use solution
  - o Low cost for companies

- Step two: B2C – To the consumers
  - o Large / growing market
  - o Able to leverage from insured to network
  - o Ambassador programs

Revenue Model

- Device sold with high markup (75% or more)
  - o Discount available with contract manufacturer for large quantities
  - o Two "mass markets"
    - ▪ Insurance companies who purchase the device
    - ▪ Consumers at large

Corporate Model

- Virtual model
  - o Focusing on our diabetic expertise to develop new products
  - o Contract manufacturing: two identified suppliers
  - o Sales through third party websites

Figure 6.3 Slides on commercialization, revenue model and corporate models.

Notes

1. Chang, Liam, & Karlin, Kirill. Deep-dive into platform biotech companies, published November 1, 2021, https://biodraft.substack.com/p/deep-dive-into-platform-biotech-companies (Last visited 10 April 2023).
2. Idem.
3. Jazz Pharmaceutical Corporate website, https://www.jazzpharma.com/about/corporate-development/ (Last visited 10 April 2023).

Chapter 7

Preparing Your Pitch – The Presentation Deck

Think back to presentations you have listened to in the past. I am sure you can recall some boring ones, with traditional PowerPoint slides that obfuscate information, rather than communicate it. It might have been that the slides were made of a "solid wall of text" (a dozen or more full sentences written in a 16 font), which the presenter promptly read like he was reading a chapter from a book. It might have been the color contrasts that made the text illegible, spelling mistakes which made the text unreadable or badly labeled graphs that were impossible to decipher.

These presentations failed, not because of the information, but rather because the people presenting it failed the *look 'n feel* of their presentation.

When presenting, the *look 'n feel* of your data is a crucial element to ensure the audience understands the message. Mishandling the presentation elements obfuscates your message, and the intended recipient will not capture the information you are attempting to share. Even worse, the recipient could interpret it completely different than the way you were trying to convey it. Presenting data with attractive and clear visuals is essential to sharing the results of your data collection and analysis.

Transforming data into eye-catching graphics and tables is not as intuitive as it may seem. While new software and platforms make the mechanical aspects of data presentation simple to master, everything from choosing the right graphics to the graphical design aspects is a unique set

DOI: 10.4324/9781003381976-7

of skills, skills I refer to as *Look 'n feel* skills. They include choices related to the look of the presentation (colors, fonts, shapes, layout, graphical) and the feel of the presentation (architecture of the graphics and tables, responsiveness and interactivity). *Look 'n feel* also includes the effort to create a consistent image across a presentation, and the overall branding image that is generated.

The next two chapters will be dedicated exclusively to communicating information to our audience. For this chapter, we be focusing on the physical aspects of the presentation: the graphics, the images and presentation tools. In the next one, we will be going more over the storytelling and how to present your presentation deck.

The key to a good graphical presentation is to select the method that best fits the data, so we will start by going through a process to choose the best way to communicate quantitative information. This will be followed by information on models for presenting qualitative data. The chapter will conclude with an overview on presentation tools that integrate quantitative and qualitative data such as slideshow tools, the Prezi platform and the use of Infographics.

7.1 Presenting Quantitative Data

The transformation of quantitative data into a comprehensible output is a necessary step to getting your message across. To do so, it is important to decide which graphical layout is the best one to demonstrate your information.

■ *Tables are used when direct access to precise numbers is necessary to support your message.*

In a *table*, information is displayed across rows and columns. A table is useful if precise individual values must be shared. For example, imagine doing a presentation to illustrate demographic data relative to your product: if sharing the precise market share percentage for each demographic segment is crucial to the presentation, the use of a table is indicated. Ideally, a table should have a clear title, include information such as the sample size and when the data were collected, and is inserted after the text that relates to it.

■ *Graphs are indicated when a relationship in data is what you want to demonstrate.*

A *graph* is data that is displayed in a visual layout, such as a pie chart, a line graph or a histogram. Visualized data allows you to demonstrate information in a relationship to another data set, along one or more axes, and is useful when you are trying to show the shape of data, its patterns and its trends. Following the previous example, if the message you wanted to circulate is how much your product is sold to each demographic segment in relation to other competitors, then the use of a chart (such as a pie chart) effectively communicates relative market size from one company to another. Ideally, a good graph will include a title, a clear scale and clear axes so the reader can quickly understand the data.

Note that if you are showcasing a single important point of information or one data point, simply use large tile or label for your data, rather than embedding it in a table, a graphic or a visual component.

7.1.1 Transforming Quantitative Data into a Graphic

If you have determined that a visual layout is needed for your information, you can identify which graphical representation is the most pertinent to communicate your message. Remember that "fancier" does not mean better. Sometimes, a simple line chart is all you need to properly convey the information you are trying to share.

A simple approach to selecting the best visual layout for your data is to identify the data relationship you want to demonstrate. Once you have decided the type of relationship you want to profile, you can use the following table (Table 7.1) to narrow the best graphic representation for your data.

As a broad example, many years ago I wrote an article for a marketing magazine.[1] Using data from a client's quinquennial survey, I wanted to explore the relationship between a product's environmental attributes and consumer purchasing patterns, as well as measuring how these patterns evolved over time. The hypothesis was that over time, environmental attributes would become more important in a consumer's decision process.

However, the data articulated a completely different story, as there was no significant difference in consumer purchasing patterns between 2009

Table 7.1 Choosing the Most Relevant Graphic Representation for Your Data According to the Type of Relationship

Relationship	Description	Graphic Representation Alternatives
Time	You are observing a single variable over time	• Use a *Line Chart* to emphasis the shape of data • Use a *Bar chart* to emphasize contrast between data sets
Ranking/Score	You are scaling items in order during a single period	• Use a *Bar Chart*, especially horizontal bars if you have a long label to place • Use a *Radar Chart* if you are comparing scores across multiple objects
Part of a whole	The items you have analyzed are part of a whole	• Use a *Pie Chart* when comparing one key item versus the whole market • Use a *Bar Chart* when comparing trends across attributes is important • Use *Stacked bars* when comparing whole categories across time
Benchmarking	Items are compared to a reference item	• Use a *Bar Chart* to compare your data to benchmarks • Use a *Line Chart* to emphasis shape of data
Frequency	You have compiled several observations per interval	• Use a *Histogram* to emphasize individual values
Correlation	You are comparing two (or more) variables across time	• Use a *Scatter Plot* to show how one attribute is affected by a second variable • Use a *Bubble* to show data from a scatter plot, emphasizing a third variable

and 2013. The importance of environmentally friendly attributes and the willingness to pay more remained at the same levels between the two surveys. As I was demonstrating multiple relationships in data (multiple time periods versus purchasing interest), I used a simple bar graph to illustrate the data. This way, the reader could clearly see that the consumer levels remained constant from one data set to the next. The use of a graph was indicated because 1) I was illustrating a trend and 2) the relationships between time and attributes were key to understanding the story (Figure 7.1).

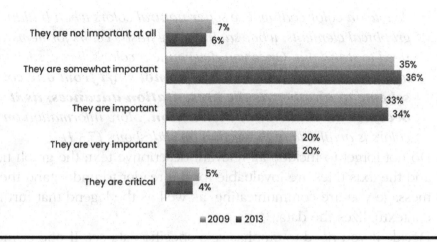

Figure 7.1   Using a bar graph to illustrate "How much influence environmentally friendly attributes have on your purchasing decision".

7.1.2 *Building a Graph*

Once you have identified which type of visual layout you wish to build, and have started design, you can use the roadmap below to be sure you are not forgetting any information.

a. *Before building the graph*
   – If you are doing a graph with mixing quantitative and qualitative scales, place the quantitative scale on the Y-axis (vertical) and the categorical scale on the X-axis (horizontal); time is especially important to place horizontally since it follows our intuitive habit of seeing time moving from left to right.
   – If you use a bar chart, your horizontal scale will most usually start at zero, but if you are using a line chart, you can narrow the scale and starting the data at a more relevant number. This will give the reader a better sense of the pattern or story you are trying to convey.
b. *Once you have built your graph*
   – Remove any distractions that have crept into your story. When building your graph, focus on the data itself, and try to remove any graphical elements that aren't helping to understand the message you are conveying. For example, the use of grid lines is not recommended unless they are useful for the reader to interpret data (e.g. the reader needs to precisely see where the data is)

- *A note on color coding: Use softer natural colors when building graphical elements, while saving more flashy colors for items you need to stand out. Once you determine a color scheme, stick to it for the entire presentation, and **do not shift from one color scheme to another as the presentation advances, as it becomes an additional distraction.** More information on colors is available in the section on slideshows (7.3.1).*
  - Do not forget to include all relevant descriptive text: the graph title and the axis titles are invaluable for the reader to understand the message you are communicating, as well as the legend that further contextualizes the data.
  - Decide if you need to emphasize a specific data set. If one element of data is more important than the others, highlight it accordingly, either using bold colors, or using a different graphical item.

7.1.3 Decision Tree Modeling

Decision tree modeling is a visual presentation model used to illustrate a decision process and their corresponding actions. It can be used to illustrate any process with choices such as a purchasing process or the impact of a marketing campaign. The top of the tree refers to the choice alternatives, followed by the decision criteria and closing with the decision outcome.

As an illustrative model, decision trees have many advantages. First, they are easy to understand, as they are a visual representation of a phenomenon, and it is very intuitive to follow. They are also very useful to enhance decision-making, or as a starting point to brainstorm. For example, it could be used to illustrate a consumer decision process. It also allows to illustrate how one data set impacts a second one.

In another example, a client in the dental supplies market was looking to offer a new service and was interested in determining how the delivery model impacted dentist interest in purchasing in relation to the price she was targeting. As such, a decision tree was modeled to help illustrate the collected data from a survey (Figure 7.2).

Using the data collected in the survey, this decision tree model was built. The choice was around the purchasing decision at a certain price point, the criterion was the delivery speed, and the outcome was the purchasing decision. As we can see, the slower the proposed delivery model, the lower the purchasing intent.

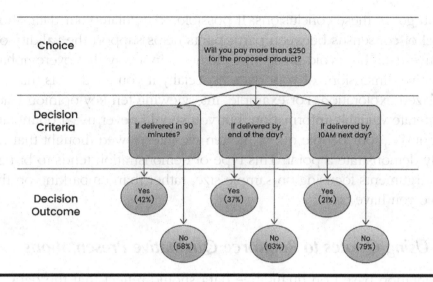

Figure 7.2 Decision tree modeling example showcasing how price, delivery model and purchase intent are interrelated.

Overall, decision trees can provide insight into decision process, enable better decision making and illustrate to the audience the data behind a specific decision.

7.2 Presenting Qualitative Data

Qualitative data differs from quantitative data during analysis, a difference that persists all the way through the presentation of data. The richness of qualitative data is descriptive in nature, which does not necessarily lend itself to graphical representations. A literature review of close to 800 scientific articles found that only 27% of articles that presented qualitative data used classical graphics to present the information.[2] Hence, presenting qualitative data focuses on the narrative and the story. To do this, we will be going over a few tools that can be used to either reinforce the presentation, or to help illustrate the collected data.

7.2.1 Overview of Presenting Qualitative Data

Presenting qualitative data means focusing on organizing the story into a narrative that makes sense to you and the readers. Hence, focus on the general themes that you have identified during data analysis, and focus on how you got to these conclusions. If possible, triangulate your data: showing the level of consensus between participants helps support the validity of qualitative data, but avoid presenting elements in a way that overemphasizes quantitative dimensions of your data, especially if you used tools that emphasized exploration. For example, interviewing ten key opinion leaders will generate valuable information, but you should never use a quantitative argument (i.e., four people out of the ten we interviewed thought that…) to formally demonstrate a point. This type of demonstration tends to bring out counterarguments focusing on sample size, rather than embarking on the narrative you have built.

7.2.2 Using Quotes to Reinforce Qualitative Presentations

A presentation based on qualitative data should reflect that the data consists of patterns found in quotes and observations. Hence, it is quite common to use direct quotes to reinforce conclusions during presentations. Quotes can also be used in reports to reinforce quantitative data, to strengthen an element that is being emphasized, or that was discovered during analysis. If you choose to directly quote someone in a presentation or a report, remember to respect confidentially and attribute quotes anonymously (unless express permission to directly quote is given).

Also, while you can use some brief quotes to illustrate a point, don't use too many in the same section of your document. Avoid over-quoting participants as a way of delivering qualitative data: quotes illustrate the results of the analysis, but do not replace it. Stringing a series of quotations without analysis makes it difficult for the reader to follow (especially if the narrative style shifts from one quotation to the next).

Finally, it is tempting to choose the most remarkable quotes from your interviews to demonstrate a point but remember that these do not necessarily represent the pattern of your data. Choose quotes that represent a concerted point of view and avoid outliers to illustrate a point.

One way to emphasize a quotation is to use a *boxed display*. Boxed displays are text framed within a box. These are used to highlight a specific narrative or quote which is important enough to be put into evidence. It emphasizes something of specific interest and helps separate a concept and a supporting quotation (Figure 7.3).

A few years ago, I completed a project with a client in South-East Asia. The project was to identify the feasibility of establishing a small bio-industrial cluster, based on regional strengths and competition. The report included several interviews with companies interested in the region, but these interviews had revealed a few potential issues. To ensure the issues were noticed by the reader and could be addressed, boxed displays (such as this one) were used carefully throughout the report, emphasizing those latent issues.

Figure 7.3   Use of a boxed display to share qualitative information (quotes).

7.2.3 *Visual Layouts to Display Qualitative Data*

The imaginative use of diagrams and schematics can be useful ways to illustrate analytical processes and findings, and to simplify more complex information. While there are many ways to visually display information in qualitative presentations, we will be emphasizing matrixes and flow charts.

7.2.3.1 *Qualitative Matrixes*

Matrixes are built by crossing two or more dimensions, variables or concepts of relevance to the topic of interest and to see how to interact. They are used to classify data across topics and demographic data or more complex illustrations of results. The advantage of building a qualitative matrix is that when you are building them, you get to understand your data further, giving you an additional level of analysis. Of course, not everyone is visually oriented, but they are an interesting way to display information.

To illustrate, let's refer to a project I did, a survey for a media client active in the Gen Y demographic space. This client was investigating purchasing patterns for health drinks. He asked participants about the last drink they had purchased, and what had led them to purchase that product over another competing brand or product. Since the objective was purely exploratory, the answers participants gave were open-ended. Once the data was codified, I created a framework with over a dozen different categories of responses. The top five answers had the most responses.

Table 7.2 Example of a Mixed Matrix (Qualitative and Quantitative Data) Showing the Relationship between Purchasing Factors and the Sex of the Respondent.

Purchasing Factor	Answers	%	Female	%	Male	%
Taste	689	68.9	336	67.2	352	70.4
Price	366	36.6	182	36.4	183	36.6
Refreshment	278	27.8	137	27.4	140	28.0
Health	271	27.1	156	31.2	114	22.8
Availability	239	23.9	117	23.4	121	24.2

To see the difference between purchasing factors (qualitative data) and the sex of the person answering the survey (quantitative data) we built a matrix (Table 7.2).

The advantage of a matrix is that it has the feel of quantitative data, and is useful to convey messages, especially when you have access to large data samples.

7.2.3.2 Flow Charts

Flow charts are diagrams used to document, analyze and illustrate a process or organizational structure using various graphical elements, connected by arrows. They help visualize what is happening, can help illustrate a problem and a possible solution. Also, they can be used to effectively communicate a process to other parties, so they understand it. In presentations, flow charts are useful to illustrate an existing process, for which a product or service resolves an issue or solves a problem. They can be a useful way to illustrate qualitative data you have gathered from interviews (how stakeholders usually perform an activity, or how they complete a transaction, for example), and triangulated with other sources of information (such as secondary research or observation).

Some conventions around flowcharts include the use of rectangles when illustrating an activity, and a diamond when illustrating a decision. Also, stay consistent throughout your flowchart: a consistent approach eliminates unnecessary distractions and lets the reader focus on the essentials. Try to keep shapes the same size as much as possible and limit the use of multiple colors. Finally, try to keep everything in one page: if it gets too crowded, it might be an indication that you have multiple processes or activities intertwined, so try splitting them up into two or more flowcharts, linking them as needed.

When building a flowchart, start by organizing the tasks in the chronological order they happen or following the hierarchical structure. Note when decision points occur, and what the various consequences are. Also, try to take note when there is a feedback loop (an activity that returns to a previous step in your flowchart). Then, do a first draft of your chart: I use a whiteboard as first drafts are often cluttered, and redundant steps may need to be removed. Once you have a stable flowchart, you can use Word or Excel to illustrate it, or a more specialized software or website if you are building something particularly complex.

To illustrate, I refer to a research project I did quite a few years ago. A CEO at an emerging start-up was interested in benchmarking personnel assigned to clinical research per product in biotechnology firms in Canada and the US. Since information in this field is usually proprietary (and not available at large), research was done through three indirect sources: review of industry literature, interviews with individuals and review of hiring history for companies of similar size and following a similar business model. The hypothesis was that by triangulating three independent sources, we would obtain accurate data.

While the results are proprietary, we can share an example of a flowchart that was built from our consolidated data. Already, this gave the CEO some perspective on organizational development and trends (Figure 7.4).

7.3 Presentation Tools

There are some interesting presentation tools that you can use to present information, especially if you are consolidating different visual layouts, models and information into a coherent story. First, we will take a moment to go over some slideshow basics, followed by an overview of visual storytelling software and infographics.

7.3.1 *Slideshow*

Slideshows are presentations consisting of information, pictures, videos and audio shared in a sequential order using an electronic device. They have the advantage of being (potentially) more attractive and interesting than a

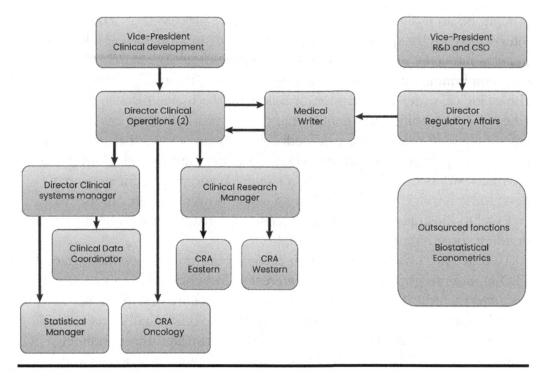

Figure 7.4 Hypothetical clinical development structure for mid-sized biotechnology company.

traditional paper document and can be exported in standardized file formats (such as a PDF) so they can be distributed at large.

While we assume that most people reading this book will be familiar with the basics of doing a slideshow, there are a few fundamentals that you should remember during construction of your deck.

a. *The least text possible:* Do not put every single word of your presentation on the PowerPoint slides. Having too much text tends to lead the audience to *read* the information, rather than *listen* to the presentation. As your audience can most likely read faster than you can speak, they will have completely read your slide long before you have shared the information.[3] They will be ready to your next slide, and most likely will not even be listening to you, missing key information or drifting away from your content.

b. *Standardize the fonts and backgrounds:* Use the same fonts (sans serif, if possible) across your entire presentation. Use one font size for your titles, and one font size for your text. A third font size could be used only if you want to specifically emphasize some key information.

Standardized fonts and backgrounds make it easier for the audience to follow without being distracted. An alternative to standardized fonts is using a font matcher (such as FontPair (http://fontpair.co/)) to find proven font combinations for both your presentations and your documents.

c. *Use the 6x6 matrix:* Synthesize your information as much as possible. Try to keep every slide following the 6x6 matrix: a maximum of 6 bullet points per slide and 6 words per bullet point. The objective is to make sure that you have synthesized your information, and you own your content beforehand. This has the added benefit that you won't be able to read your slides mechanically.

d. *Test your presentation onsite:* If possible, arrive early and test your presentation on the computer provided, as well as testing the projector: There is nothing more destabilizing than having technical issue or a formatting glitch midway through a presentation. If the content is not too sensitive or confidential, send your presentation ahead and ask the recipient to test the presentation on their hardware to make sure there are no issues.

e. *Color management:* Colors can increase participants' interest, but incorrect use can create distraction or make the presentation ineligible. As a rule of thumb, try to keep cooling colors for the background of your slides (such as blue and green), and warmer colors (such as orange and red) for objects in the foreground. Also, be consistent throughout your presentation: once a color scheme is chosen, keep it for the remainder of the presentation unless there is a specific reason for shifting (i.e., calling attention to a specific element). There are many online tools that can be used to manage colors and choose ones that match best. Personally, I am a big fan of Coolors (https://coolors.co/).

7.3.1.1 Slideshow Software

There are a lot of alternative slideshow software available on the market. Here are some of the most popular or interesting ones out there:

■ *PowerPoint* is by far the most widespread slideshow software. It is well known by many users, is simple to use and can be quickly set up for a presentation (most laptops have PowerPoint installed on them). It also allows you to export slides into PDF format, making it easy to share content in a secure format. It comes with many templates pre-installed, but due to the prevalence of PowerPoint, many audience members will be familiar with most of the standard templates so if you use them, you won't be winning any points for originality.

- *Google Slides* is a powerful alternative to PowerPoint. Like PowerPoint, it lets you integrate most format files into your presentation and integrates well in Google Software Suite. The main strength of Google Slides is its collaborative aspect: it allows multiple users to access and update slides simultaneously. If you are working in a collaborative online environment, Google Slides might be the alternative for you.

 **Philippe weighs in on getting ready for your presentation:** Bring screen copies of your presentation so you can present even if you are without internet access. Indeed, with the emergence of the cloud, many companies put their presentation online, but if you show up to a place without Wi-Fi or internet, you will come off as unprepared.

Why not always have a hardcover version of your slideshow? It happens often that after a day of intense presentation, your laptop or tablet is out of battery. It is best to always have a quick version to show especially in a trade show context with one-to-one meetings at conferences like BIO.

7.3.2 *Visual Storytelling Software*

Visual storytelling software is an alternative presentation platform that enables interactive presentations. Rather than being confined within the chronological order of a slideshow, presenters can adjust their presentation to the audience's interest and questions.

Billed as a conversational presentation tool, visual storytelling is variation of the traditional top-down presentation. The rationale behind these platforms is that by the time your presentation starts, the attendees will already have researched a lot of information on you and your concept from your website and web presence, and they may already know what their gaps in knowledge and their interests are. They might already know very well what information they need to decide. Rather than being a keeper of information, it is now the attendees who direct the presentation flow by asking questions and expressing interests. The objective is to have a natural conversation and focus the presentation on what the participants are interested to find out.

One of the leaders in this space is Prezi (prezi.com), which has the additional advantage of being usable offline using an APP and is available on Android and desktop versions.

7.3.3 *Infographics*

While classical methods of presenting data are well-known and recognized by the community, there is a limited visual appeal to them. Also, there is very little that is distinctive about them. One of the tools to emerge in this space is the use of *Infographics* to present data.

An infographic is a graphical visual display of information and data used so the information can be assimilated quickly by the reader in a pleasant manner. Hence, once you have found or created interesting and reliable data, infographics can be a useful tool to produce engaging visuals and create interest.

The advantage of using Infographics' is that they show overall context in a clear and visually appealing way. Immediately, the viewer can easily understand what the document is about as well as see how each data set is related to each other. Infographics are appealing because it is easier for people to process images than it is to process text. Also, they are a modern way of presenting otherwise traditional data and are portable, as you can easily insert them into a business plan, a webpage, a physical printout for distribution or even post them across social media platforms.

Combining elements of table presentations with graphical elements, infographics allow the user to communicate messages in an attractive way to the audience. But to do so, the user will have to move beyond the traditional software suite. There are several interesting platforms that possess powerful capabilities to build infographics such as Canva Infographic Maker (www.canva.com/create/infographics/) and Venngage (venngage.com).

The limits of infographics are that is that the researcher is limited in the complexity of the information he can share, and that abusive use of infographics can lead to situations akin to "PowerPoint poisoning", i.e., the overload of information presented in such a disjointed method that it bores and loses the recipient's interest. As such, try to limit the amount of text on your infographic, privileging images and statistics instead.

One of the other main issues with infographics is data accuracy. If you have one incorrect data point on your infographic, it will quickly impact your overall credibility of the other data and the infographic in general, so make sure your data is completely accurate. Also try to limit the number of colors you use on a single infographic: too many colors make it difficult for readers to understand the information you are trying to share with them. Finally, while you might have a lot of great information to share, be careful not to overload on content: excessive information will impact user readability. Making users wait for the page to load, or having to scroll down will reduce the overall effectiveness of your infographic. If you have too much content, consider splitting the information into two separate documents, rather than trying to cram everything into a single document.

As an example, I have included an infographic I prepared for Ancon Medical many years ago, a client active in the early cancer detection space. This is based on the information found in his business plan as well as different market research reports I prepared for them. Hence, I was able to build an infographic that conveys information on the importance of early cancer detection as well as showcasing some of the key relevant market data we had found that supported the company's vision (Figure 7.5).

7.4 Slideshow Framework: The 10/20/30 Rule

The 10/20/30 rule of presentation was proposed by Guy Kawasaki, a Silicon Valley venture capitalist specialized in marketing. Having seen numerous pitches and presentations, he noticed that entrepreneurs were concentrating on unimportant content and losing focus. Since then, he has been advocating the 10/20/30 Framework. The objective is for the presenter to synthesize his corporate presentation down to the essentials: "A PowerPoint presentation should have **ten slides**, last no more than **twenty minutes**, and contain **no font smaller than thirty points**".[4] A complete description of his approach is available on his blog. The ten slides that he deems essential are

1. The problem: What is the problem you have identified
2. Your solution: How are you fixing the problem
3. Business model: What is your organization's business model

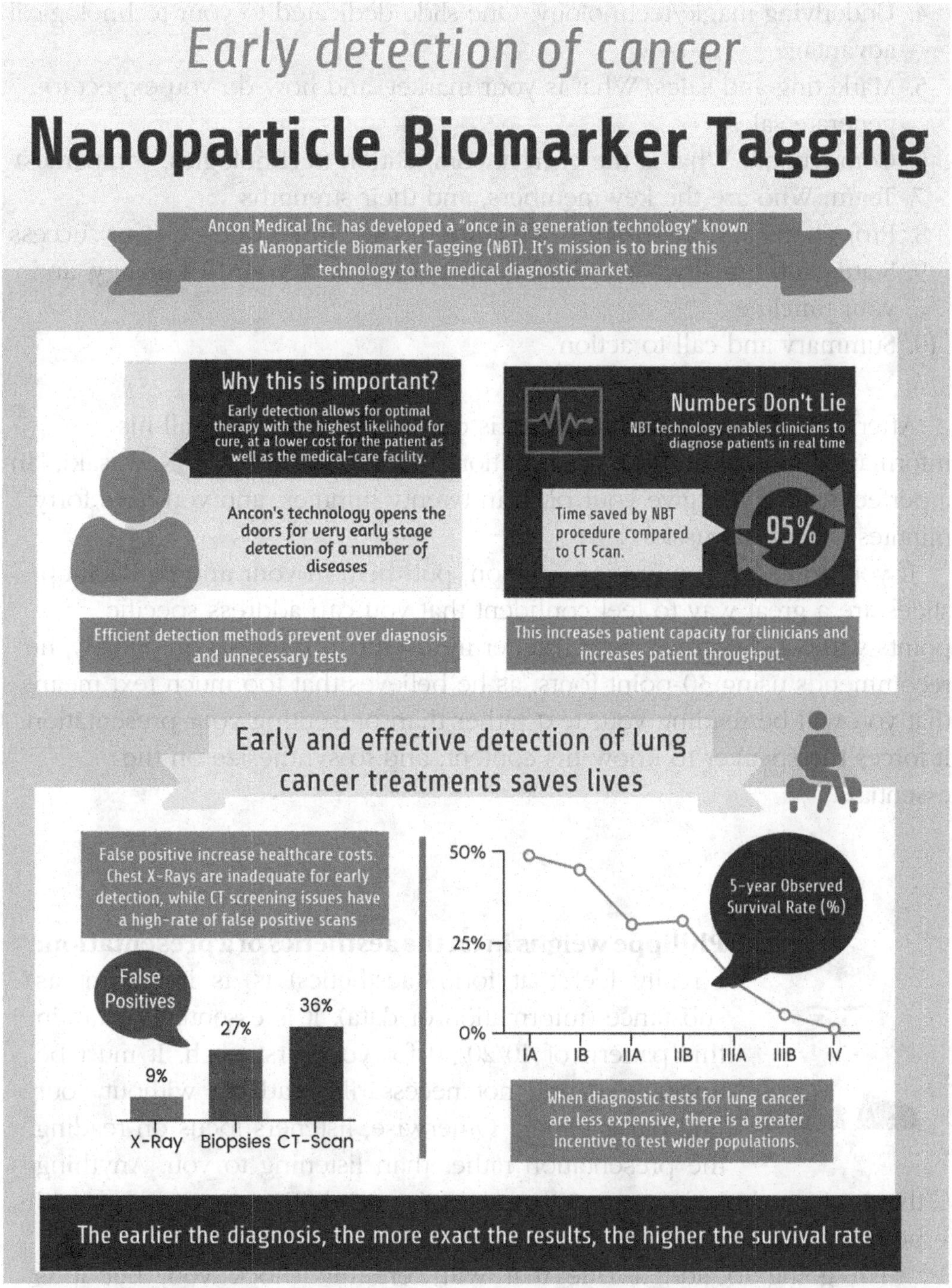

Figure 7.5   **Example of an infographic for a company in the early cancer diagnostic space.**

4. Underlying magic/technology: One slide dedicated to your technological advantage
5. Marketing and sales. What is your market, and how do you expect to generate sales
6. Competition: What is the current competition (technologies, companies)
7. Team: Who are the key members, and their strengths
8. Projections and milestones: What are the key milestones to your success
9. Status and timeline: What is the current status of your technology, and your timeline
10. Summary and call to action

Afterwards, the onus of the effort is on the presenter to fit all his information in a 20-minutes presentation. As mentioned by M. Kawasaki, "In a perfect world, you give your pitch in twenty minutes, and you have forty minutes left for discussion".

If you do feel you more information, put them in your annex. Back-up slides are a great way to feel confident that you can address specific points without drowning your listener in too much information. Finally, he recommends using 30-point fonts, as he believes that too much text means that you will be reading your text rather than presenting your presentation. It forces the speaker to know his content, and to synthesize on the essentials.

Philippe weighs in on the aesthetics of a presentation: I really feel that form (aesthetics) is as important as substance (information or data). It is essential to stay in this pattern of 10/20/30 for your first pitch. It must be uncluttered and not necessarily readable without your oral participation. Otherwise, listeners focus on reading the presentation rather than listening to you. Anything that is of a technical nature or that gives rise to additional requests should be dealt with in the slides in the appendix or in backup.

I'm going to add a rule that will certainly shock you, but it is essential that a teenager be able to understand your messages and your slideshow. It is essential that you can explain complicated or scientific or technical concepts to any average person. Pitch to one of your children or someone who has nothing to do with the subject. The person will give you unexpected feedback and improve your presentation (often by simplifying).

As we will see in Chapter 9, the teams of investment funds see a lot of passing of files and presentations. It is necessary to avoid that the red buzzer is pressed. And the slightest mistake will eventually trigger the "red buzzer". In America's Got Talent, the red buzzer is a red button located on each of the judges' tables. If judge wants the performer to stop, then they press the buzzer.

7.5 A Few Tips on Preparing Content

In general, we believe you should be able to address the three following points before your presentation.

7.5.1 Ask Yourself the Right Questions

As you prepare your presentation, answer the following questions, as they might come up during your pitch:

- Why am I looking for this type of financing (VC) rather than another?
- Do I have sufficient notoriety in my field?
- How am I going to generate performance for future shareholders and give them the practical possibility of benefiting from it (exit strategy, liquidity)?

7.5.2 Gain the Support of "Ambassadors"

Prescribers, key opinion leaders, customers... Being able to show that you have the confidence of people recognized in the field... This gives your project credibility! Prior to the launch of your fundraiser, it is therefore important to raise awareness of your project among the people likely to support it.

Also, don't forget to invest in a press relation network. Good visibility will create a significant interest.

7.5.3 *Don't Neglect the Operation Side of Your Start-Up*

Financial projections are used to show potential investors your vision and your desire for development, as well as the potential of the project; they are therefore essential. However, they must be based on a pragmatic operational version of this vision, which you must be able to explain in order to convince of the feasibility and realism of the project. Too many entrepreneurs get stuck in the strategic side of their project, neglecting how they will operationalize their vision in real-world operations.

7.6 Closing Remarks – Simplifying Technology

I remember attending a client's pitch a few years back. He had prepared his slide deck and felt quite confident about his presentation. He had invited me to attend to a pitch he was making as to get my opinion on potential investors and to give him feedback on his presentation skills. He was quite proud of his technology. So proud, in fact, that his presentation focused almost exclusively on the technical innovation and prowess of this technology, with very little focus on the market, commercialization plan and business model. When he finished, one of the investors candidly asked, "What are you selling exactly?" After that presentation, I must admit I was as confused as the investor.

On multiple occasions, I have been asked by clients to review and comment on their "pitch deck". One thing I constantly point out is that these slide decks are far too technically focused, with many slides dedicated to the science, but addressing the audience from such a specific angle that they are ineligible to the average person. While your technology is undeniably a key element of a company's unique selling proposition, if the audience is unable to understand the technology, it will quickly become white noise.

While some individuals possess exceptional research skills, creating paradigm-shifting technologies and redefining markets, they must remember that not every person in the audience might possess the same level of skill. Hence, it is possible that the message never reaches the recipient as he is just unable to understand it. It is important to take note of the audience's knowledge level and customize the message to match their level of understanding. You might have to explain what you feel are basic terms and concepts, so be ready to describe them. If possible, develop stories, examples and parallels to simplify complex technologies.

Philippe weighs in on your key message: Be clear about the value proposition! Unfortunately, after half an hour of presentation, it is common to wonder what exactly the company is doing … Ideally, you should be able to start by describing the business with a sentence like this:

We [the company] are a [business] that offers [a solution] to [target customer segments], bringing them [a unique and differentiated profit] Our business model is [monetization of service or product]. We expect [3-year turnover] with a net result (or gross margin) of x%

Finally, in the initial meetings, some emphasis must be given to the market opportunity that the technology provides, and how the company intends to approach the market to commercialize the technology. Later, more time will be made available during due diligence to demonstrate scientific and technical validity of the technology.

Notes

1. Denault, Jean-François. 2014. How much do environmental attributes influence purchasing patterns? Marketing Magazine. https://www.marketingmag.com.au/hubs-c/how-much-do-environmental-attributes-influence-purchasing-patterns/#.U4MYzfldXL8 (Accessed 11 May 2023)
2. Verdinelli, S, and Scagnoli, N. 2013. Data display in qualitative research. *International Journal of Qualitative Methods, 12*(1), 359–381. https://doi.org/10.1177/160940691301200117.
3. If you speak faster than attendees can read your presentation, you possibly have another presentation issue on your hand.
4. More information on the 10/20/30 rule is available at http://guykawasaki.com/the_102030_rule/ (Accessed 14 June 2023).

Chapter 8

Preparing Your Pitch – The Story

The objective of a pitch is to convince investors that you have an opportunity they cannot miss. The investor will look at the extent to which your project can enable him to achieve his performance objectives. It's up to you to demonstrate that you are well-aligned with their needs and that the entrepreneurial project you are proposing corresponds well to the investment fund's timing and investment strategy.

In summary, if during your pitch you are thinking **"give me money!"**, don't forget the investor will be thinking **"give me some performance!"**

Over the years we've seen hundreds of pitch decks, some of them helped raise millions of dollars and some of them were thrown in the bin. We still find ourselves quite surprised to see entrepreneurs underestimate how important it is to create a bulletproof and attractive pitch deck. In the next few pages, we will be going over the five things to remember before building your pitch as well as seven components of the perfect pitch to make your pitch stand out from the rest.

 DOI: 10.4324/9781003381976-8

Jean-François weighs in on the three What's of storytelling: Presenting your company to an investor is about taking them through the three **what's**: 1) what happened (the problem), 2) so what (why is this important) and 3) what now (how are we solving this). In life, stories create recognizable patterns. Be it a movie we are watching, a book we are reading or even a presentation we listening to, our brains look for patterns. They look for continuity. If you are able to write a compelling story, you will find your audience that much more invested in your pitch.

8.1 Five Things to Remember before You Start Writing Your Pitch Deck

8.1.1 Give It Time!

Many entrepreneurs rush off to make their pitch without first taking the time to make sure they have the best tools (Figure 8.1). Like our entrepreneur below in Figure 8.1, they are too busy with the end results, not taking into consideration the importance of the tool and preparation before getting to their task. But truth be told, it takes time to prepare a powerful presentation! You will have to work on and rework your pitch to make it effective. Each sentence must be thought out and validated. You have no room for improvisation or hesitation.

Figure 8.1   Feedback is key in an entrepreneurial project.

If you're an entrepreneur looking for funding, you know how important and involved the pitch process can be. Designing a compelling pitch deck and identifying an effective hook is just the start. In fact, testing your pitch to see if it resonates with investors is the next step. Time is also the opportunity to get feedback on your project.

8.1.2 Practice Your Skills!

The best way to prepare a strong pitch is to train repeatedly and in front of different audiences, especially with people who are not familiar with your sector of activity or your company. Practice your pitch with your partners, mentors, friends, and make the adjustments needed. Two major components to pay specific attention to:

- Ensure you master your presentation. You must be comfortable when you present, and able to adapt on presentation day if necessary. In the end, **you must be able to make your presentation without your deck and know its content by heart.**
- Once you have started practicing pitching the deck with others, confirm that the content is clear, concise, dynamic, credible, convincing. Ask them THE question: ***Would you invest in my company?*** Why? What convinced you, or turned you sour on the opportunity?

8.1.3 Tailor Your Pitch to the Audience

Even if the framework of the pitch does not change from one presentation to the next, it is always beneficial to customize and adapt it to the listeners, as this will enhance their interest. Before starting to write your pitch, you should do some basic research and answer the right questions:

- What are the VC's expectations?
- What types of investments have they already made?
- What are their investment criteria?
- What is the fund's exit horizon (liquidity)?
- What are their passions? Their journey?

Finally, take a moment to identify why you would want to have the VC you are presenting to as an investor over someone else. Beyond the financial aspect, what can they offer you?

8.1.4 Be Committed

Fundraising is a time-consuming process. You must be prepared for it as much as possible and the leader must be able to devote most of his time to it. Part-time CEOs are fine for the first few steps of fundraising, but once you get to VC capital, you must be able to demonstrate you are committed and have faith in the innovation and company you are building. You are all in.

Hence, your personal commitment (as well as the team's commitment) is essential to inspire confidence because it constitutes an indicator of the motivation of the founding team and of their confidence in the project.

8.1.5 Highlight Your Team

A project does not rest solely on the CEO: the team counts, and it must be built in line with the project. Present yourself as much as possible as a team (two or three people) by highlighting your complementarity, the role of each, and "who ultimately decides". This last point is especially important as investors need the reassurance of knowing there is a "pilot on the plane".

 **Philippe weighs in on the key factors that influence whether you will be able to successfully raise venture capital:** Getting the first investor on board is extremely difficult. The three key factors that influence whether you will be able to successfully raise venture capital are your ideas, your network and your pitch. Having a well-designed and structured pitch deck is one of the keys to raising capital.

8.2 The Seven Components of the Perfect Pitch

8.2.1 Be Concise

You have little time and every word counts.

8.2.2 Be Clear

Your public must understand immediately what you are explaining, which means that you must forgo jargon, overly technical terms and confusing terms. This is sometimes difficult for the entrepreneur passionate about his subject. To address this, pitch in front of an uninformed audience, and look for confused expressions. Then, re-work your pitch to address these complex topics.

8.2.3 Be Specific and Transparent

You must give confidence, and not let people think that you have things to hide or that you are not ready to discuss in total transparency. For example, avoid being vague and using words that elicit suspicion. Except in exceptional cases, you shouldn't have to hide information. Also remember that if this information comes up later in the due diligence, it could derail your whole relationship with the VC.

The investor expects a significant capital gain. Hence, you have to illustrate "how are you going to make money?" Hence, it is important that you can show a precise calendar of operations: from the moment the money is brought in, what will it be used for? By what timeline?

Also, keep your feet on the ground in terms of valuation, and don't just think in terms of percentage of capital. Ultimately, it is better to hold a smaller share of a good company than a large share of a worthless company. Display a valuation in line with the achievements demonstrating the progress of the project.

8.2.4 Have a Dynamic and Impactful Attitude that Resonates with People

You must keep in mind that your audience has many important things to do, and their time is valuable. Your pitch must be short, surprising and captivating. You will have to make your mark and "awaken" your audience.

8.2.5 Pay Attention to Your Speed

Be careful not to speak too fast or your interlocutor will understand nothing. Take your time (think about 150 words per minute).

To find your natural flow, do the following exercise: take a stopwatch and read a text for one minute. Then count the number of words you just read.

This is a theoretical rhythm, as silences are an important component during a pitch. Furthermore, you can (and will probably) be cut during your speech. Finally, take into account that your speed will certainly increase with the pressure of the presentation. If you have half an hour for a presentation, try to allocate about half of your presentation to content and the other half can be filled with questions and answers. This has the added benefit of gauging participants' interest in your presentation.

 **Jf weighs in on questions during the presentation:** I usually ask the audience to hold questions until the end of a slide: Some specialists advocate that the audience should wait until the end of the presentation to ask questions, but there are some issues with this approach. The main problem is that by the end of the presentation, your audience will often have forgotten their question. Worse, some will struggle to remember their question all through your presentation, focusing on their question and not listening to new information as you share it. Others suggest that people in the audience ask questions as they occur to the point of "interrupting" the presenter when they have questions. This is not something I recommend as well, as it constantly interrupts the flow of your presentation, and many times, questions that are asked in the middle of a slide are answered by the end of the slide.

To minimize issues with the flow of your presentation (and to limit interruptions), you can ask people to wait until you finish each slide or pause occasionally to ask your audience "Are there any questions up to this point?"

8.2.6 When Writing Your Presentation, Follow the "7 S" Theory

1. **Simplicity:** keep it simple and clear,
2. **Story:** tell a story,
3. **Slides:** Master the slides. Images are important to mark the audience,
4. **Surprise:** Don't hesitate to surprise the audience to keep their attention, and to be creative,
5. **Senses:** Touch the audience. Make it aware and make sure it feels concerned. Don't hesitate to appeal to its sense of humor,
6. **Say again:** don't hesitate to repeat the keywords or catchphrase,

7. **Summary:** At the end of the presentation, summarize the outline of the project to refresh the public's memory and connect the different parts.

8.2.7 Sell Yourself and the Future Performance of Your Business

The pitch allows you to make a first impression on the managers, and the investor will evaluate several points:

■ How cohesive is the team? Is there a clear decision-maker? (*They will ask questions directly if the presentation does not make this point clear*)
■ Do they seem convinced? (*If you are not confident when doing a pitch for an investment, how can you be confident when comes time to sell your project?*)
■ Is their approach professional? (*Professionals associate with other professionals. Once a VC invests in your company, they become tied to it, and they certainly will not want to be associated with a poorly organized company*)
■ Do they resist pressure? (*You might be tested by some of the audience, asking hard questions where the answer is not as important as how you answer it*)
■ Are they good salespeople? (*They will assess your ability to convince them and your future partners or customers*)
■ Are they presenting something realistic? (*One of the most important elements is always to have validated the project management triptic (timing/cost/scope) by checking with key contacts (subcontractors and suppliers for example)*)

Philippe weighs in on the value of validated information: I remember being particularly impressed by a team with cofounders who had prepared their whole project impressively. They were developing a therapeutic. All the important costs (preclinical tests, production costs, regulations costs and so on) had been validated by at least three quotes from the top five suppliers of the industry. Trust was instantly created. Unsurprisingly, it is a company that has been particularly successful in following its financial projections and its development roadmap.

8.3 Closing Words

Simply put, avoid bullshit as much as possible and go straight to the point, be yourself, convincing, pragmatic and above all humble. If everyone finds themselves around a table to listen to you, it is necessary that there is an interest.

Also, remember that while you are in a loop where you are saying to yourself "give us money", the limited partners of the funds (the investors who have invested in the fund) are also in their own loop, with a mantra of "give us financial performance".

Who says fundraising does not necessarily mean guaranteed success? In fact, 20% to 50% of the start-ups supported by a VC end up going bankrupt. While high, it is two to four times lower than unsupported start-ups, which both testify to the difficulty of lasting in increasingly volatile markets, as well as the importance of working with the right investors.

Hence, try to choose your investors according to what they can bring you in addition to their money (such as their notoriety, experience and network) and especially according to the affectio societatis.[1]

If you are concluding a deal with an investor, make sure you have found a good match and start on a good basis. Remember that this is a "mariage à durée déterminée" and that you will be working together for the next few years. Resentments are always bad in the long term and never in the interest of the company!

Note

1. The affectio societatis designates the will to associate, the common
   will between several natural or legal persons to associate. It constitutes
   a characteristic element of the contract of partnership. In French law
   and Quebec law, there is no legal definition of the will to associate but
   jurisprudence has therefore defined it.

Chapter 9

Identifying the Right Person to Pitch To

 Investors have been injecting capital at an accelerated rate into the private equity market for almost 20 years. In Europe for instance, between 2003 and 2020, the amount invested increased from 0.2 to 1.9 billion euro.[1] This is a trend we have also seen in the US, as invested capital in the health sector has grown from 26.3 billion USD in 2017 to reach 59 billion USD in 2022.[2]

The consequences of this abundance of capital, in a low interest-rate environment, are directly visible: there is more competition between funds to invest on available assets which has led to growth in valuation multiples (a trend accentuated by the COVID-19 pandemic in specific sectors such as technology and health). In 2021, because of the post-COVID-19 recovery, the average valuation multiple in Europe reached an all-time high of 12.7, up from 9.7 five years ago. Even more, in France, the record of 13.43 in a single valuation multiple was crushed! The expansion of multiples was a determining factor in the sector's performance in Europe.[3]

However, this rapid growth does not mean that your journey to secure investments will be easier. There are two essential elements to successfully start your investment journey: First, as we mentioned earlier, it is essential to have a clean and precise pitch. Second, it is essential to be able to find the right person to pitch to.

In the next few pages, we will give you an overall feel of the investment process, and then dive deeper on how to identify the right person to talk to,

DOI: 10.4324/9781003381976-9

go over three steps on how to identify the right investor for your start-up in Section 9.3, as well as the single step to identify the wrong investor for your start-up in Section 9.4.

9.1 A Few Basics About VCs that You Have to Know

For the moment, let's assume the pitch is clear.

Before you find the right person to talk to, it is very important to understand the inner workings of a venture capital company. During the investment process, there is always a strong collaboration between the analysts who carry out the analysis of investment opportunities and the partners who present new investment cases to an investment committee. **It is important to be in touch with both parties.**

Also, before contacting a fund, it is crucial to evaluate the sector of activity the fund focuses on and the maturity of the technologies it targets, as well as making sure the fund is active in its "investment period". It is therefore important to understand the fund's strategy and its investment strategy.

Once you are in touch, it is crucial to hone your pitch, because you only have one chance to make it to the next step.

Investors have an extremely stringent selection process. As we will see later on, a "normal" Venture Capital (VC) fund might hear an initial 3,000 pitches every year and have to whittle down these applications for two to three investments. Unless they love your team, your technology and the economic model, the selection process for a VC is particularly tough and selective. So, any opportunity to say "no" will be seized.

As an entrepreneur, your objective will be to always have a next step with your point of contact, and thus avoid at all costs the "no", until a final decision is a "yes".

"The best way to convince somebody, is to put yourself in their shoes and understand their needs". So, before answering the question "Who should I talk to", let us stress "what" a VC is, what is the purpose of a VC and how do VCs make decisions.

9.2 The Basics of the Venture Capital Fund

When a private equity firm creates a private equity fund, it forms a "Limited Partnership". The firm puts out a call for investors to contribute to a pool of capital that will be used to invest in private companies that

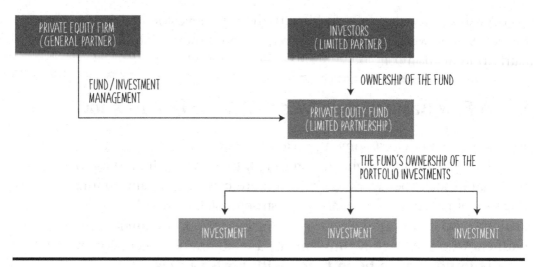

Figure 9.1   Private Equity Fund structuration.

fit within a predetermined investment strategy. By contributing, investors become "Limited Partners" (LPs) of the fund. The private equity firm is the "General Partner" of the fund and responsible for managing investments. The goal of a private equity fund is almost always to eventually sell the stake in the company for a high return on equity. The strategy that funds follow to reach that goal depends on the fund and the sub-asset class it targets (Figure 9.1).

The purpose of a venture capital fund is to responsibly generate returns for LPs by funding innovation and advising entrepreneurs (which is also called VE Returns, or Venture Equity Returns). LPs are those that provide capital which VCs invest through a fund. LPs are generally not involved in the operations of the funds and have limited liability. The most important LPs can be represented on the advisory committee of the investment fund.

A venture capital fund is an investment vehicle with its own rules. These define, among other things, the proper functioning of the fund, as well as the investment strategy of the fund. (See Figure 9.2.)

VCs operating in the same segment, strategy or territory are in competition for the best investment opportunities, and frequently target the same opportunities. This can have the indirect impact of raising acquisition prices and reducing investor performance. In order not to give in to a flocking effect (where all VCs seem to target the same

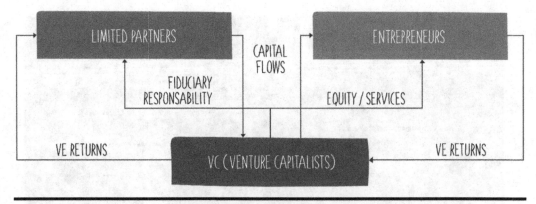

Figure 9.2   Dynamics and relationships between LPs, VCs, and start-ups.

companies, raising valuations), VCs strive to innovate and offer LP novelties in terms of:

a. **Sectorial specialization**: operating beyond traditional sectors, some teams have developed the expertise to create funds dedicated to sectors that are exponentially growing (such as impact investments, artificial intelligence, big data or fintech) or traditionally resilient activities (such as infrastructure or energy).

b. **Geographic specialization**: these funds look search for opportunities new areas with strong growth potential such Asia, Latin America and Africa.

c. **Strategic specialization**: Some funds focus on a specific business cycle, targeting companies in the early development stage, those raising a series B, leveraged buy-out (LBO[4]), mezzanine debt, fund of funds and so on.

Each investor has its own selection criteria and its preferred targets. Although there are many investors, the number of investors that you are a target for is quite small. When identifying investors to contact, it is important to ensure that you fit into their selection criteria (Figure 9.3). This background research is indispensable.

As you can see in the cycle, there are many different investment vehicles. Currently, the most popular vehicles are detailed in Table 9.1.

Some people might refer you to existing investor databases and directories. While these tools are a good starting point, remember that the world of private equity changes very quickly. Some of the funds, at

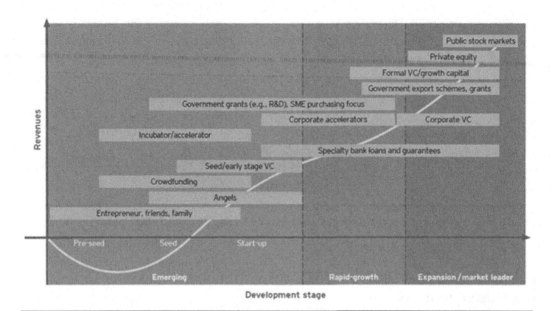

Figure 9.3 Investment cycle and different financing vehicles.

Table 9.1 Most Popular Investment Vehicles

Crowdfunding or participative financing	Raising small amounts of capital from multiple individuals to finance a new business venture. In life science, Syndicate Room, Wiseed, Seedmatch and Companio are relatively popular.
Business angels	High net-worth private individuals, who directly invest in new and growing private businesses in the hope of getting high return.
Family offices	Entities created by families to act as investors on their behalf, with a structure similar to a VC, but a long-term ROI perspective.
Institutional investor	Fund that makes investments on behalf of someone else such as pension funds, mutual funds and insurance companies.
Corporate venture	Created by large companies, generally operating in their own sector of activity. The best way to identify them is by identifying the big players in your sector of activity, and then investigate if they have a fund or an investment structure.

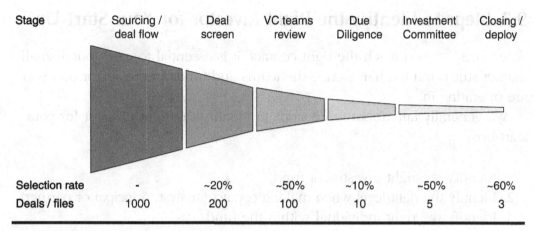

Stage	Sourcing / deal flow	Deal screen	VC teams review	Due Diligence	Investment Committee	Closing / deploy
Selection rate	-	~20%	~50%	~10%	~50%	~60%
Deals / files	1000	200	100	10	5	3

Figure 9.4 Deal flow process for VCs.

the time the data was collected might have had a specific strategy, but since the data was collected, might not have done any financing or had limited activity in the identified category. Some might have reoriented themselves toward other targets, other might not have any capital to invest. On the other hand, many active funds are not listed, because they are not members of the main associations or have come into place since the preparation of the directory. As such, we have found that these tools might jumpstart your research, but these tools do not replace your own direct investigation.

It is also important to understand the deal flow processing within an investment team. For example, and according to the selection stringency of the investment team, 1000 companies might be investigated (sourcing phase) which would lead to about three to four investments (deploy phase). (See Figure 9.4.)

So, the first task a VC faces is connecting with start-ups that are looking for funding—a process known in the industry as generating deal flow. For each deal a VC firm eventually closes, the firm considers, on average, 200 opportunities. A total of 25% to 50% of those opportunities will lead to a meeting with management; 20 will be reviewed at a partner meeting; five to ten will proceed to due diligence; five will move on to the negotiation of a term sheet with the start-up; and only two to three will be funded. From a VC perspective, a typical qualified deal takes between 70 and 100 days (about three and a half months) to close, and firms spend an average of 120 hours (about five days) on due diligence during that period, making calls to an average of ten references.

9.3 Steps to Identify the Right Investor for Your Start-Up

To ensure that you reach the right contact, it is essential to carry out a small market study and to characterize the teams and funds in the sector that you are operating in.

We generally talk about three steps to identify the right investor for your start-up:

1. Identify the right investment fund,
2. Identify the right level when making contact (analyst, principal or partner),
3. Identify the right individual within the fund.

9.3.1 Step 1 – Identify the Right Investment Fund

There are multiple criteria to evaluate when you are researching funds. Some are more fundamental, while some may require a lot more research. As such, you could do this as a two-tier evaluation. In the first stage, you should investigate the three following basic criteria.

- The investment stage of the fund: What is the maturity of the projects that the fund finances: Is it seed funding, series A, B or a mix of maturities?
- The sectors financed by the fund: Which sectorial activity does the fund specialize in? Is it B2B technologies, consumer products, fintech, software, biotech, AI…? Contacting a fund that doesn't finance your sector is as absurd as a waiter bringing you a Fanta instead of a Coke.
- The portfolio of the fund: What is the type of company which has already been financed by the fund? Has your sector already been financed by the fund or even better, can there be synergies with other start-ups?

During your research, you should verify if there is a match between these elements and your start-up. If not, move on, don't waste time. But if your start-up matches with these elements, you should start the second step and do a deeper dive and investigate the eight following criteria;

1. **The VCs preferred stage of intervention:** is the private equity firm involved in seed funding, company creation, development, or takeover? Do not contact a fund that does not intervene in the development stage of your start-up,

2. **The status of the private equity firm**: is it national (state owned), semi-national or private? The investment criteria for entry into the capital may vary depending on the nature of the company that holds the fund,

3. **The minimum and maximum investment amounts granted:** there is no need to contact a fund whose minimum amount is much higher than your financing needs, companies seldom change their investments amount targets based on opportunities,

4. **Sectors of intervention:** some funds specialize in specific sectors of activity, in which they have considerable expertise and a network. For other investors, although not specialized, they often have privileged sectors,

5. **Geographical coverage:** does the structures operate on a regional, national, European or global scale? Funds usually have a well-defined geographic footprint, although following COVID-19, we have seen more funds take a look at deal outside their traditional sector of activity,

6. **Their type of stakeholding:** Do they take minority or majority stakes on their deals, or does this vary from one deal to the next?

7. **Their experience:** consult the websites of investment funds, which will usually include their investment policy. You will be able to look at the list of their participations and validate their targets (amounts, stage of development, sector, etc.),

8. **Their activity:** is the investor currently active? Does he invest? According to what criteria?

 **Jean-François weighs in on identifying the right VC:** We talked earlier in Chapter 5 about customer personas as a tool to develop your commercialization strategy. Building VC personas can also be a very useful tool, both in identifying and practicing your pitch, anticipating issues and questions. In this case, some of the characteristics would include maturity level they usually invest in, preferred technologies, level of investment and engagement levels.

9.3.2 Step 2 – Identify the Right Level when Making Contact

For some background, here is the organizational chart of a typical VC fund (Figure 9.5):

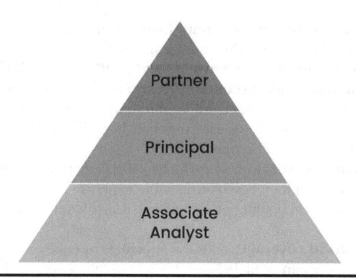

Figure 9.5   The classic organization chart of a VC fund illustrating the three main decision-makers.

As we can see, there are three main decision-makers you could decide to contact.

The top two levels, the partner and the principal can commit the VC company to an investment and are also associated in the fund. They are also remunerated on the performance of the investments. It might be intuitive, but worth mentioning: the higher the investor's position in the organization chart, the more impact the decision-maker has in the decision to invest. It seems logical to favor contacting a partner. Even if intuitively, we tend to contact the partner directly, in practice it is a little more complicated as it will depend on the bandwidth she or he has available. Without introduction to the partner, making contact can be very complicated.

Starting with a first meeting with an analyst is often a mandatory step in all funds. Without an introduction to a partner, we recommend that you favor this approach.

Overall, each person on the investment team is important in a fund's investment decision, but it remains a priority to convince partner or principal if you hope to be financed.

One final tip as we wrap-up this section. In general, startups tend to target investors who know the sector and have a track record. One of the mistakes to avoid is to contact a fund that has already invested in a project very close to yours. In general, unless there is a concentration strategy in a particular sector, an investor will prefer to avoid investing in a company that could compete or create conflicts of interest with its existing portfolio. This may also have a benefit known as "risk dilution" in the fund portfolio.

 **Jean-François weighs in on identifying VCs interest in your technology:** To add to this important point, the best strategy is to identify a fund that might have an interest in a technology, without presenting a project that completely overlaps an existing investment. For example, imagine a fund that just invested in a diagnostic technology targeting cognitive diseases. Another company with a similar device would not be an interesting target for this fund, but a company developing something complementary in the same space (e.g., a digital therapeutic targeting cognitive diseases) could be of interest as it could add value to an existing investment.

9.3.3 Step 3 – Identify the Right Individual within the Fund

Once you have identified the fund and the level you want to target, it is time to identify the right individual. To recognize the right individual for your project, you should identify the individual investors affinities within a fund (most partners will develop a specific specialty within each fund) and see which companies this person has already financed (these are also often listed on the web page). Some other clues that you should keep an eye out are conferences the partner has spoken at (especially the content and topic), any professional and business associations they are volunteering/member of as well as any board memberships they might have.

Identify the right investor from the start, the one who will have the most affinity with your project. This necessarily maximizes your chances of being invited to present. He will better understand your project and will be more inclined to become an ambassador for your start-up within the fund to promote an investment. We found that the best way to approach the individual identified is to attend to conference or an event where the individual is possibly participating to shortly introduce yourself. Then, quickly present your company and opportunity. Either he will tell you to send him a short deck, or he will refer you to the right person in the company to contact.

Finally, a quick word on sending a cold email (without any third-party introduction). While some funds are open to this type of approach, for most funds, cold emails usually have limited impact, because the bandwidth of investors and more particularly of partners is very limited. If you have identified a fund, you believe is a good fit, better to invest some time to either arrange an introduction or meet them at a networking event.

9.4 Talking to the Wrong Investor

Private equity brings together all the players who invest in equity in companies. The actors are very varied; it is important to understand their specificities and their differences to better target those whom you will choose to address, according to your needs and the situation of the company. The best way to contact the wrong investor is to disregard the category of investor in relation to their stage of development (seed, creation, development, IPO) and in relation to their investment strategy (sector, stage of development, etc.).

Here are some tips and especially the main mistakes to avoid when raising funds.

■ If your project does not meet the fundraising criteria of the fund, don't exhaust yourself going around the place, and identify other financing solutions; the first being the customers which are great non-dilutive financing source!

■ There is no point in sending your business plan or your executive summary to all investors, you must target them! It's a small world where everyone knows each other, and this kind of strategy is seen as a crippling lack of discernment.

■ Avoid common pitfalls to boost the appeal of your project (overselling) such as exaggerating sales volumes and turnover!

■ Avoid giving investors the impression that their money is going to be used to remunerate the project promoters more than to develop the company!

■ Avoid saying you have letters of interest or term sheets, as you begin your fundraising process when you do not have one. Investor cycles are always long and remember the large number of pitches (deal flow) that a private equity company receives can be anywhere between 1,000 and 3,000 per year.

■ Avoid showing a forecast that is already out of date and no longer relevant! As the months pass, it must be updated regularly according to the results…

■ Negotiations are long… Anticipate and avoid doing them "with your back to the wall" and "finding the knife to your throat".

■ Be careful not to overvalue a contribution in kind to mechanically "inflate" the company's valuation! This scares away investors, because the value of a company is not calculated according to the number of unpaid hours you have spent there (even if they are important and have allowed the progress of the project…).

9.5 Closing Words

The choice of the investor and the right person to pitch to depends on several criteria: the stage of intervention, the status of the private equity firm, the investment amounts granted, sectors of intervention, the geographical coverage as well as their type of stake holding.

Investors' experience is key. Do not hesitate to contact companies in their portfolios, consult the websites of investment funds, they will inform you about their investment policy. You will be able to look at the list of their participations and validate their targets (amounts, stage of development, sector, etc.). As soon as you have identified the right investor, the one who will have the most affinity with your project, we believe that the best way to approach the person identified is to attend a conference and be able to contact them directly.

For each investor, several sources make it possible to validate your target list: their website, local economic players, who may know them, exchanges with entrepreneurs who have already raised funds and made this journey, articles that you will be able to read in the press on fundraisers of the same type as yours, and also indicator published by fundraiser.

Notes

1. Club des Operating Partners, France Invest & Bain & Company. Les équipes opérationnelles.
   Nouveaux catalyseurs de valeur pour les fonds d'investissement, April 2022. https://www.franceinvest.eu/wp-content/uploads/2022/04/Livre-Blanc-du-Club-Operating-Partners-de-FI.pdf (Last visited 10 April 2023).
2. Venture capital in the USA, Dealroom.co, 21st of January 2023, https://dealroom.co/guides/venture-capital-in-the-usa (Last visited 22 April 2023).
3. France invest – Invest Europe, https://www.franceinvest.eu/en/ (Last visited 14 June 2023).
4. A leveraged buyout (LBO) is the acquisition of another company using a significant amount of borrowed money (bonds or loans) to meet the cost of acquisition.

Chapter 10

The Venture Capitalist's Rulebook to Investing

 How VCs make investment decisions is a very systemic process. It is organized in successive well-established phases. Each of these can be very intensive and long, depending on the actions of either party (the start-up or the investor).

Investment funds spend a lot of time managing their investments and analyzing several projects in parallel before making an investment. They are very busy and can only devote a short time to each company that does a first pitch. Negotiations can drag on when you have several investors around the table, each with different interests and constraints to reconcile. It is therefore important to anticipate these deadlines, and to prepare the fundraising effort as well as possible, so as to reduce all the steps that hinge on your preparation.

Often companies hear that they will go through a due diligence process. It's important to understand that "due diligence" is ultimately the final step before your company is presented to the decision-making investment committee who makes the final decision. But before getting to the due diligence, there are many important stages: Sourcing, Deal Screen and Partner's review. Following due diligence, you have another two steps: the review by the investment committee and deployment.

The deal flow process is a funnel where hundreds of prospective companies go in but only a small percentage are actually getting an investment. Some VC teams review between 100 to 1,000 companies for

 DOI: 10.4324/9781003381976-10

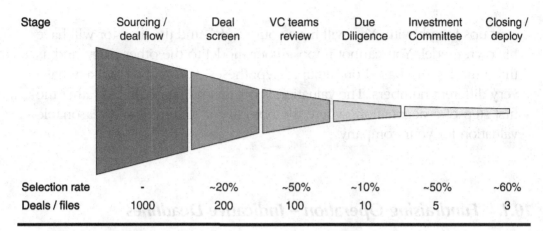

Stage	Sourcing / deal flow	Deal screen	VC teams review	Due Diligence	Investment Committee	Closing / deploy
Selection rate	-	~20%	~50%	~10%	~50%	~60%
Deals / files	1000	200	100	10	5	3

Figure 10.1   Illustrating the deal flow, from the venture capitalist's perspective.

every investment they make. This number depends on the size of the VC firm (asset under management) and the number of funds they are managing. As a reminder, take a moment to review a typical deal-flow (Figure 10.1).

As we see in this funnel, while a typical venture capitalist might see a high number of opportunities for his fund (over 1,000), he will usually end up investing in less than 1% of these opportunities. In the next few pages, we will go over each of these steps, and share what happens behind closed doors.

10.1 Preparing for the Venture Selection Process

Convincing investors to finance your project cannot be improvised. Before contacting a fund, it is therefore important to prepare. We have prepared this handy checklist before first starting your fundraising effort. (See Table 10.1.)

Having these elements on hand will reassure the investors of the potential of your start-up.

Jf weighs in on your company valuation: This is a topic many start-ups struggle with as they structure their company. In some projects we did models using discounted cash flows, in others we document past equivalent transactions. But there is one thing I always say to the start-ups I work with: You will have your model, and the investor will have his own model. You cannot impose your model to the other party, and as these models are based on multiple hypotheses, you will most likely have very different numbers. The valuation is not an ending point, but rather the first step of a negotiation, where the objective is to agree on a reasonable valuation for your company.

10.1.1 Fundraising Operation – Indicative Deadlines

The process can vary a lot from one investment cycle to the next, but in general, a venture capital investment will follow the following timetable:

■ First step: Preparation of documents and first pitch: (three months)
■ Second stage: Meetings and exchanging information on the project: (three to six months)

Table 10.1   Start-Up's Checklist before Starting Fundraising

- **Update your website:** This is an important step that is often overlooked yet can provide a lot of value. If an investor is meeting with you, his first reflex will be to do a quick web search of your company. Having a clean and well-documented website ensures you make a good impression.
- **Identify the experts** who will accompany you in your fundraising effort. Business leaders who have experience in fundraising often give good advice.
- **Prepare important fundraising documents** such as:
  – Your executive summary: summary of the pitch to present the project and allows investors to assess the feasibility, credibility and potential of the project rapidly.
  – Oral presentation slideshow or "slide show" or "pitch": you must have prepared it and tested it with as many "competent" people as possible.
  – Business plan: detailed presentation with financial statements and projections,
  – Additional documents for "due diligence": these can include your I.P. portfolio documentation, and current certifications/regulatory documentation.
- **Have a demonstration to share** (on video or on hand)
- **Have customer (or potential customer) references** that can be contacted (or letters of intent, if you are pre-commercialization)
- **Get an idea of the valuation of the company** (with or without the help of experts),
- **Get ready to build a dataroom to share your documents.** Tools exist but opt for a simple and effective solution

■ Third stage: Negotiation and finalization of the operation: (three to five months)

There are multiple factors that can influence these delays:

■ Availability of team members on both sides of the transaction.
■ Responsiveness of investors and ability of entrepreneurs to provide the necessary documents and information to investors. Some investors often take longer in responding as this allows them to see how the management team is doing, whether other investors are interested, and if the company is reaching some of its early predicted metrics.
■ Number of trips during the negotiation.

As a start-up, preparation and fast responses can go a long way to shortening the investment cycle.

10.2 The Deal Flow – Step by Step

10.2.1 Stage 1: Deal Sourcing

The first stage in the deal flow process is broadly referred to as "deal sourcing" or "deal origination" in venture capital. This is the process of finding appropriate leads and bringing them to our attention. How do we source deals? We start by sourcing deals through our personal networks and referrals, although some dealmakers also use direct deal-sourcing tactics. For instance, some funds have a specific sourcing focus (such as geographic region, academic origin technologies, specifics technology maturity stage.). When this occurs, VCs develop or use an existent network or ecosystem (such as incubators, accelerators, laboratories, tech-transfer offices and venture partners) to originate deal opportunities. For the start-up, being part of the ecosystem increases its visibility, increasing the chance a VC will notice it and contact it, or be receptive following a cold call. As an entrepreneur, you have three channels to be included in the deal flow of a VC firm:

■ First possibility: you submit your file on the VC company contact form. This is the unsolicited flow or **"inbound deal flow"**. This is the least efficient way to get visibility. Unless the pitch contains exceptional metrics (growth, monthly recurring revenue, key publication, key industrial partnership, etc.), the investor has no one "pledging" for the pitch deck, meaning no one has vetted or pushed the opportunity. Very few deals come from this channel.

■ Second <u>possibility</u>: you request one of your contacts, who is part of an Associate's or Partner's networks, to recommend you to them.

This is the recommendation or **"referral"**. The person can come from a professional network such as an incubator, a group of business angels or even a university. It can also be from a personal connection. This is the second type of contact. The recommendation is more effective because the Associate or the Partner has confidence in the contact who recommended the file to them, will take greater care in reviewing the pitch deck.

■ Third <u>possibility</u>: The last option (which I believe is the best) is for the Associate or the Partner to meet you directly at an event such as a conference or trade show.

This is the requested flow or **"outbound deal flow"**. This is the best option, because the VC comes to meet the founder, and already has an opinion of the potential of the startup.

Once a company is sourced, it usually goes to screening. When this happens, the Associate entrusts his analyst with analyzing the pitch deck and supporting documents such as the business plan or traction data (revenue, customers, cohorts).

The analyst will scrutinize the various crucial points of the file: the team, the product, the market, the comparative advantage. If your opportunity seems interesting, the analyst will confirm with the Associate, and they will organize a first meeting with you to present your pitch. This is usually done in a face-to-face meeting.

10.2.2 Stage 2: Deal Screening

It's at this step of the deal flow process that the VCs hold an initial meeting with the candidates they decided they'd like to learn more about. The goal is to collect specific information that will help the firm determine whether the company is a good investment fit.

During the first meeting, you will meet the Associate and his Analyst. You will present your company's pitch deck and answer the questions they might have. At this point, your company will also be assigned a dedicated lead or point of contact at the VC firm. After the initial meeting, the lead uses the information that has been collected to compare potential deals and select the most competitive opportunities.

At this stage, the file has already passed a selection process based on secondary research. If there is an organized meeting, it is because the company presents a profile compatible with the investment strategy of the fund or the dedicated team.

After the meeting, the Associate and the Analyst perform an internal review. They will compare your deal with other similar deals in the fund's portfolio. Out of an average of ten deals and founders screened during the week, the Associate will select his top 1 or 2 opportunities (which hopefully includes your company!). These top opportunities will be presented during the weekly Partner Review.

10.2.3 Stage 3: Partner Review

Next, Associates present the competitive opportunities that they've selected to the firm's Partners. Of these leads, only 10-15% of the startups presented in this meeting are selected to move forward to the next stage, making it the most narrow part of the deal flow process funnel.

The Partner review is a 2 to 4 hour meeting, during which the investment team will present their best new deal opportunities of the week to the Partners. The lead Associate on your file (the person responsible for its sourcing) will present their deal for 10–20 minutes. After that, Partners (including the Partner who works with the Associate), will share their feedback on the deal and ask questions.

There are three possible outcomes to the Associate's proposal:

1. The lead Partner and the other Partners are not convinced, and they decide to pass on the investment opportunity. The Associate then contacts you to announce the end of the evaluation process for your company.
2. The lead Partner and the other Partners are uncertain and prefer to have additional information before deciding. The Associate then passes on the questions to you, who must provide the required data. The Associate will be able to present the answers at the next Partner review.
3. The Partner and the other Partners are convinced by the Associate and your startup. The Associate can start due diligence (10–15% of the startups presented pass the Partner Review). In addition, the Partner, interested in your startup, will be declared lead Partner on the deal. In

this case, the Associate then contacts the entrepreneur to announce that he/she will carry out due diligence and that the Partner will be the lead partner (the partner in charge of the file) on his/her deal.

10.2.4 Stage 4: Due Diligence

This step is where the VC firm does a deep dive into each potential opportunity, including doing a competitive analysis, interviewing customers, suppliers, partners, and trying the product or service. Completing the due diligence phase can take up to ten weeks, sometimes even more.

At that time, the Associate, under the supervision of the lead Partner, proceeds to perform an in-depth evaluation of the startup from many different angles. The Associate will seek to understand more deeply the business model, the market environment, the competition in place and the exploitable growth paths. During this period, the Associate will sometimes follow up with the Founder for new data, but the purpose of the assessment is for the Associate to answer the question: would I invest in this business opportunity or not?

After evaluating the start-up, the Associate performs due diligence. This is the complete analysis of the start-up. The purpose of the due diligence will be to confirm the information collected during the evaluation. During this period of four to six weeks (which can extend up to ten weeks if necessary), the Associate performs various exercises. The Associate will set up a data room covering several subjects (governance, legal, finance, product, intellectual property, etc.) and analyzes various elements (sales pipeline, acquisition funnels, funding).

The Associate will also carry out objective exercises such as trying out the demo of the product of the start-up, perform reference checks on the founder(s) and the company as well as performing several question-and-answer sessions. If significant elements contradict the preliminary analyses or if the figures presented do not turn out to be in line with reality, the deal will most likely be abandoned.

All this work accomplished will be summarized in a **memorandum for the investment committee.** It's at this meeting that we will decide whether or not the deal will be carried out concretely. During this time, you and your team will have to hold the helm, avoid any major issues, otherwise, the Partners might take this as an opportunity to stop the evaluation process and leave.

On average, 10 to 15% of files accepted for Partner Review pass the due diligence stage.

10.2.5 Stage 5: Investment Committee

An investment committee is made up of the firm's partners, industry experts and other important and experienced professionals. We can find operational or industry experts, or even investors invited for the occasion (especially if they are invited to be secondary investors on the deal). This is the team that reviews and discusses all of the information the lead associate and partner captured during the due diligence process, and ultimately makes the decision about whether or not to invest.

The investment committee meets for approximately 1 to 2 hours after which the decisions about whether to invest or not are recorded.

During the meeting, the lead partner will present the memo, the analysis and insights drawn from the due diligence. This is followed by a question and answer session with the committee and a decision is made. Note that in some cases, the company is invited to present a deck to the investment committee for 30 minutes. At this step, the Partner and the Associate will help by indicating the key elements to prepare. They can also review ("proofread") the documents and give their feedback for an optimal presentation. This usually depends on the people involved and their available bandwidths.

There are three possible outcomes for the Investment Committee:

a. The committee is not convinced of the opportunity, and it is rejected.
b. The committee is hesitant and asks for new due diligence before deciding. The lead Partner and the Associate will have to review the decision and plan an ad hoc meeting to decide if they wish to pursue due diligence.
c. The lead Partner convinces the committee to invest, and the deal goes into the closing phase.

In the event of a favorable decision to continue and to enter more detailed negotiations, the investor writes a Letter of Intent (LOI) or a Term Sheet (TS).

This document specifies the conditions of its intervention: valuation of the company, financial conditions, legal conditions and main clauses which will be found in the shareholders' agreement. If you agree to counter-sign this document, a new negotiation begins to finalize valuation and draw up the investment deal.

If you have several investors around the table, the "lead investor" will be the one who will orchestrate the discussions. The lead investor is often either the investor who is the most organized, the one who is investing the highest amount or the one who agrees to take a position first to generate interest from the others.

10.2.6 Stage 6: Deploy Capital

Following the investment committee's decision, the lead partner and his fund propose a contract formalizing the VC company's participation in the startup: the term sheet. The term sheet is a legal and commercial document of five to ten pages formalizing the agreement between the company and the investor.

The lead partner and the founder or co-founders will negotiate the final terms of the deal. Among these terms, there are points relating to fundraising such as the pre-money value of the startup or the volume of the investment round, but also post-raising clauses such as liquidation preferences or anti-dilution protection. The goal is to be as precise as possible in the drafting of the terms, to prepare the drafting of the large format: the shareholders' agreement.

Once the negotiation is over, the final term sheet is written, and the proposal is sent to the founder or the founders validation. Once the term sheet has been validated, then the shareholder agreement is drafted by a specialized legal firm. The founder or founders of a start-up obtain their financing against cessation of a share to the Venture company.

Then follows what this whole operation is for, the funds are transferred to the start-up. The associate is responsible for archiving the file and associated due diligence. With the venture capital investment being held over a maximum of four to ten years, the lead Partner and the Associate will work together with the entrepreneur or the founders to allow the start-up to evolve and achieve its objectives. Less than 1% of startups contacting a fund end up being funded.

The deal has officially been closed! The final term sheet is signed, and the funds are wired to the startup.

10.3 Due Diligence: How to Judge the Situation of a Company

A key step in the investment process occurring downstream of the letter of intent, due diligence is decisive for any equity investment. This verification of the elements announced by the company makes it possible to limit (unpleasant) surprises. Several elements are investigated as part of the due diligence: the financials, the legal due diligence, the technology, the management team, the environment and the commercial strategy.

The financial investigation allows the investor to identify the company's main current expenditure items, its need for operating financing as well as its key assets and its commitments vis-à-vis its debtors. A financial due diligence has a dual perspective. It seeks to first validate the investor's valuation assumptions by looking at historical performance, if available, and concluding on whether it is consistent with forecasts. It allows to identify financial uncertainties and exposures which could disrupt the business, or result in additional costs.

Legal due diligence is conducted in order to identify the main disputes to which the company could be exposed. Most of the elements will be provided in a data room with, for example, all of the company's contracts (employment contracts, insurance contracts, premises rental contracts, loan contracts, distribution contracts, etc.), warranties, pending and potential lawsuits. The objective is to understand all of the company's obligations. A legal due diligence is typically completed by an attorney who specializes in due diligence investigations. A legal team will prepare a legal opinion based upon all of the gathered factual information.

Technology due diligence is one of the key aspects as innovation is one of the main drivers of startups. It is essential to understand the company's research and development strategy. The teams will ask to provide a list of contacts or directly contact professionals in the sector (consultants, manufacturers, experts, scientists, etc.) to validate the veracity of the elements provided by the entrepreneur on its technology. The objective is to fully understand the positioning of your technology on the one hand and the various milestones necessary for the proper development of your product or service. This is yet another way of properly modeling the risks and putting them into perspective with the company's technological development roadmap. It is also the opportunity to fully understand and characterize the technology positioning in the market to be addressed.

The management team is another item where it is essential to carry out an in-depth audit. Interviews are organized with the key members of the team on the one hand, as well as key partners. Due diligence will also be carried out on the founders of the company and the members of the executive committee. This consists of asking for references, people with whom they have been working with their whole career. Visits to the company offices and labs are organized to get to know each other better.

It is also an opportunity to participate in demonstrations. While this step might seem trivial, it is an essential part of due diligence as it may lead to some unforeseen discoveries.

A few years ago, I was completing a due diligence report for a company we were looking to invest in. During a visit to their offices, we visited the laboratory. After a demonstration on the device, I found myself chatting with a member of the team. I noticed a similar device to the one he had just demonstrated and asked him what this second device was. He told me that it was an obsolete version of the new model under development but that this outdated version was already being commercialized by a third party located in China. After a bit more research, our firm found that the manager of this second company was a member of the co-founder's family. This information, which had never been mentioned at any point of the due diligence, was shocking. You can easily understand that after this visit the trust was completely broken, and we stopped the due diligence process.

The environment in which the company is immersed is a determining factor in its future success. The business environment is defined as its relation to everything outside: technology, nature of products, customers and competitors, other organizations, political and economic climate and so on. The business environment includes all the elements that are external to the business, which are not directly subject to the control of the business, but which are likely to exert an influence on it and on its ability to achieve the goals it has set for itself. Studying the global (macro-economic) market in which the company is positioned will help to define and understand its mission, vision, strategy and business model. Indeed, this will make it possible to detect trends in the sector, changes in user behavior and consumer needs, but also to know regulatory constraints, market players and changes in market players.

Also, as you may be aware of, the ESG (environmental, social and governance) movement has become increasingly present in investment decisions. It's not some catchy phrase anymore. ESG is real and every company must adjust to succeed. ESG has been integrated into VC investment process for several years. So, ESG principles should be integrated at the early stage of a company's development so it will be ready and prepared for an ESG due diligence. It is no longer enough for the investor to include in an agreement clause that the company is ESG compliant. A simple checkbox or questionnaire won't be enough to ensure due diligence. ESG factors play an important role in the initial investment decisions, as well as the ongoing development of companies after investment. As a European investor, I often used the guidelines proposed by InvestEurope, found on its website (https://www.investeurope.eu/invest-europe-esg-reporting-guidelines/).

IP due diligence: IP assets are increasingly important for businesses to the extent that, for some, the IP rights owned by a company have a higher value than all the tangible assets. Just as the entrepreneur wants to know if a truck is ready to operate or if the building they are planning to buy belongs to the person who claims to be the owner, an investor needs to know about the status of the IP assets of the company that they intend to acquire or invest in. Moreover, according to tax and accounting rules, said assets can be included in the company's balance sheets and thus, its correct identification, protection and evaluation will have a positive impact on the overall value of the business.

As we saw earlier in Chapter 4, IP due diligence (DD) often needs a Freedom to Operate (FTO) analysis, to identify potential barriers or limitations to the manufacture and commercialization of the technology to be acquired, by identifying relevant third-party IP rights and assessing the possibility of infringing these IP rights by the prototype product or proposed process. The FTO is an element that can ultimately be requested as part of a due diligence by an investment fund.

Commercial strategy: Finally, the company's business strategy (marketing and sales) will be scrutinized in order to estimate the company's growth potential by analyzing the sales force and the various distribution channels. The evaluating team will most likely ask to be provided a list of contacts or will directly contact professionals in the sector (consultants, manufacturers, experts, etc.) to check the feasibility of the business plan, to understand the market and better understand the needs of the market. It is also an opportunity to fully understand and characterize competition and market risks.

Jf weighs in on your commercial strategy: This is where a lot of the market research will come into play, because the more documented and formal you were when you prepared your documentation, the more you will be able to share to the VC. For example, if you did a product-market fit, where you talked to 20 doctors, and now have documented this list of interviewees, with coordinates so the VC can independently validate, it will give a lot of value to your market projection. Hence, this might help accelerate some of the due diligence, as some of the groundwork has been done following a structured process. Don't neglect to document your research efforts, and if you are doing an interview, always end the discussion by asking if you can share the person's coordinates to potential investors and partners.

As part of fundraising and due diligence, you will be required to share strategic information. So make sure that the security and confidentiality of your documentation is guaranteed, choosing a secure **data room** tool and setting up a non-disclosure agreement (NDA) with all potential investors. Particularly during fundraising, organized and secure legal documentation will save you precious time and peace of mind. It should never be forgotten: as long as an agreement is not ratified by the shareholders, it is not finalized.

10.4 Negotiations

10.4.1 Financial Negotiations

Financial negotiations include several aspects, but center around the valuation[1] of the company. There are complex methods for calculating this. You can be accompanied and build your own valuation model, but the final valuation of the company will be the result of the negotiation between both parties.

When talking about valuation, it is important to distinguish the "pre-money value" from the "post-money value" of the company: The company has a so-called pre-money value before the contribution of the investor, and once this contribution has been made, the new value, known as "post-money", is equal to the pre-money value plus the investor's contribution. This gives us the following equation: post-money value = pre-money value + amount invested.

Valuation is a delicate phase. The figures announced by the entrepreneur are sometimes far from the estimates of the investors. It is then a question of finding a compromise. Nonetheless, it is important that at the end of the negotiations, everyone finds their benefit. Otherwise, resentments will always come out and will be detrimental to the company.

Another critical point that should not be overlooked is that future valuation rounds will not be constrained to the valuation displayed at the time of investors invested. Entrepreneurs frequently forget that the valuation given when the check is signed is not always the one that will be "used" in the end. Two other equally crucial elements come into play:

There are **valuation readjustment mechanisms** that allow the valuation to change over time and can make it easier to reach a compromise. Recapitalization systems can be put in place by investors (in the form of convertible bonds with variable parity, warrants, sale of shares, etc.). They can allow the management team to increase its share of capital if the results reach the forecast. It is strongly advised to prefer this mode of incentive, which also has the merit of reassuring investors, to systems of dilution in the event of failure or delay, which are often "double penalties".

Apart from the financial entry conditions, the conditions indicated in the clauses of the shareholders' agreement are also very important, and the exit conditions. A mistake would be to think only in terms of dilution. Other elements must be considered in the negotiation "package", such as freedom of management, the clauses of the agreement in terms of rights and control (double voting rights for example, etc.) and the clauses of the pact providing for the exit.

10.4.2 Legal Negotiations

Once both parties have agreed, the investors' entry into the company's capital is validated and the letter of intent is signed, the entrance of the capital amounts to the "marriage" and "living together" processes must be organized. This is the role of the shareholders/partners agreement.

The shareholder/partner's agreement is the document that governs the life of future partners or shareholders. Its objective is to provide guidance and direction for as many situations as possible during the life of the company, and the challenge is to find a balance between the rights of investors, the rights of historical shareholders and the interests of the company. This negotiation stage can also be quite long.

The shareholder agreement is a confidential legal document (unlike the bylaws), which is officially signed on the closing day.[2]

This new stage of negotiation is long, so do not rush it: many companies, in a hurry to get capital, shorten the negotiation cycle in the hopes of getting the funds quickly. This just sets them up for problems down the road.

10.5 What You Need to Know about the Shareholder Agreement

A shareholder agreement is a contract, (akin to a "marriage contract"), between all the partners, and is intended to remain secret. Since it is a contract, it is less authoritative than the bylaws of the company, and therefore great care must be taken to make sure the clauses do not contradict the bylaws. In addition to being a legal contract, it is a moral contract that lays the foundation of the relationship, and the parties engage in negotiations to ensure everyone's willingness to agree and find a balance.

The shareholder agreement manages the life of the company. It sets out all the rules for proper functioning, at least those which are not in the bylaws. The objective is to envisage and anticipate potential roadblocks and issues, and to foresee the best way to resolve them.

It also includes mechanisms in case the shareholders want to "divorce". You must be conscious that one day you might "separate" from your investor, and that everyone has an interest in ensuring that this happens in the best conditions. The shareholder agreement thus helps to manage the "break-up", which will avoid any "fighting" in case of disagreement,

The shareholder agreement evolves over time. A shareholder agreement (SHA) agreement is signed for a given time. Any new investor into the capital (or one that leaves/exits) implies a new SHA. In fact, any meaningful change in the life of the company may call for a change of the SHA.

A SHA is not a substitute for common sense, but it is always difficult to have common sense during conflict. Having a document written in peaceful times helps immensely and can allow everyone to keep their minds during a drawn-out conflict.

10.5.1 The Clauses of the Shareholder Agreement and Their Issues

In this section, we will not go into details of all the clauses, but rather those that come up the most during negotiations. Without being thorough, here are the main clauses that exist and their usefulness: those that stabilize the capital, those that prepare the exit and those that protect the investors and founding partners. It is especially important to understand their issues.

10.5.1.1 Clauses Relating to the Development of Capital

The **capital allocation clause** is a clause that allows parties to plan how the capital of the company transform in case specific events occur. For example, if a founding partner leaves the company before a certain time, if specific results are reached, or if there is a second fundraising round. This clause is especially important to ensure that the founders' team remains united and focused on concrete objectives.

The **pre-emption clause** allows one or more partners to have priority over the shares of an outgoing partner or resulting from a capital increase.

10.5.1.2 The Commitment Clauses of the Founders

Since the project is based on the founders and the initial management team, these clauses are intended to ensure their loyalty to the company:

The **non-competition clause** prevents a partner from performing similar functions with a competitor.

The **exclusivity clause** prevents a partner from having another activity, even a non-competitive one, to make sure that they devote their full time to the project. This clause can be adjusted to allow limited time spent on another specified activity,

There is also a group of clauses known as good leaver/bad leaver clauses. A **Bad leaver** clause is a penalty clause in case of serious misconduct, triggering the resale of shares of the "outgoing partner" at a discounted value. Meanwhile a **good leaver** clause is used in the case of the departure of a partner where the retaliatory measures would not apply or would be limited (only because of the possible prejudice for the development of the company). These clauses are usually paired with a sustainability period: this is the period during which the founders commit to remain in with the company.

10.5.1.3 Clauses Protecting the Interests of Investors in the Event of a Transaction

An **Exclusion or forced redemption** clause allows excluding a partner because of certain events, agreed beforehand, and at a price or methodology determined in advance (for example, if there is a hasty departure). The paid-up units are distributed among the other shareholders according to the other clauses (in particular the pre-emption clause),

A **Revision clause** (better known as a ratchet clause) aims to protect investors by limiting the risks associated with a subsequent valuation that is lower than the one investor invested in or one that is significantly lower than expected. By implementing this clause, the investor recovers additional shares to cover the shortfall.

10.5.1.4 Investor Exit Clauses

The objective of an investor is to make an investment that, if possible, lasts as long as possible and that allows him to realize significant added value. The different ways to take out the capital are the subject of hard negotiation. A number of clauses can be included to address the topic.

First, there is the **Joint Exit Clause** (also called the tag-along clause). In this clause, the majority shareholder pledges not to sell its shares without giving the minority the opportunity to do so equally and on the same terms. It allows each partner to sell its shares in the event of a meaningful change in control of the company. This clause makes it possible to protect minority shareholders by forcing the large shareholders (and especially the founders) to find buyers for the "small investors" at the time they themselves sell.

There is also the **drag-along clause**. Basically, it is easier to sell a company in a single block rather than splitting it into pieces (thus making it more "liquid"). As such, it is sometimes deemed necessary to bind all shareholders together to force the sale of all the shares in a single transaction. Thus, minority shareholders must transfer their shares to the third-party transferee chosen by the majority, in order to facilitate the transfer of 100% of the capital, preventing blocking by minority shareholders. Negotiations are usually necessary to determine what percentage of shareholders is necessary to activate the clause.

Also important is the **priority exit clause**, which entitles one of the shareholders to assign its shares in priority over the other shareholders. It allows one or more partners to sell their shares first in the case of a purchase of the company. This is interesting for investors who have exit constraints when the buyer buys only part of the capital.

The **Global Forced Forward Clause** (buy or sell) clause is quite specific and allows a shareholder to offer his shares to another at a certain price. If the latter refuses, he is obligated to sell his own shares at the price indicated to the partner who proposed the deal. On the investors' exit horizon, it forces managers to buy back their units or to look for a buyer for the investors' units.

As for the **alternative offer clause** (also known as the Russian Roulette clause or the American clause), it allows the shareholder who wishes to sell his shares to another shareholder at a fixed price he has determined. If the latter refuses, the seller will have to buy the shareholders' shares at the proposed price and will then become buyer of the new securities.

Other clauses include the **Deadlock clause** which defines the conditions for resolving a situation in which the company is unable to function and the **Sales Mandate clause** which allows, at a certain point (often when the investors' exit horizon is exceeded), to force the sale of the company by entrusting the buyer's search to a third party, who is then mandated to sell the company.

10.6 Questions and Answers (Q&A) during Exchanges with Investors

There is still a long way to go after pitching your project, and you still must perform well during the crucial Q&A minutes.

If you were able to show yourself in your best light during the pitch, the next phase allows you to display how you react when you are challenged. The investor will try to better understand your personality, your leadership, your skills and your ability to adapt. And so, beyond the answers themselves, how you answer is just as important. Here are some questions to expect and that you should prepare to answer.

- **What is the "ultimate" goal for you and the co-founders?**
  You should always be ready to answer this question. When you are asked this question, investors want to know if your goal is to commercialize the technology, to license it, to sell the company before getting to commercialization, to make a capital investment, to create a unicorn, to enrich yourself both professionally and financially or to change the world. When he asks this question, the investor is aligning his fund's philosophy with your own.
- **What is the state of the competition? How do you differentiate yourself? What stops a competitor from doing the same as you?**
  You are a new entrant in a market where companies are already active. How will you manage to position yourself differently and especially to prevent others from catching up with you? Be clear about

the competitors, direct and indirect... Do not say that you have no competitors, it is not credible. Your solution is always an alternative solution to an existing situation: if you are automating a process, people are doing it manually. If you are digitizing information, your clients are using a pen and paper. Think in terms of process, not necessarily technology.

■ **How long is the sales cycle? Who are the decision-makers? With which interlocutors are you in contact?**

Now is the time to show that you have a "boots on the ground" vision and that you know the target to which you want to sell to. You must show how the customer makes his decision, who makes it, what time is to be expected between the first contact and the sale. The more concrete your answers, the more you can demonstrate your real-world knowledge and your ability to commercialize the technology.

■ **In concrete terms, how are you going to sell?**

If the fundraising is to be used to set up a sales force, you will need to explain which sales profiles will be recruited, and on which realization rates you are basing your assumptions. For a start-up company, it is recognized that the best salesperson is always one of the founders of the company and that at the fundraising stage, it is not necessary to add a lot of salespeople,

■ **How will you acquire your clientele? For what cost?**

Show that you have tested several approaches and that you know which ones to use when the time comes. Also know what it will cost you to acquire a new customer,

■ **On what assumptions are the forecast figures based?**

Keep in mind the main indicators of your costed business plan, without having to read the slides. It shows that you have mastered them and that they really correspond to a vision. For each of them, be able to explain how you calculated them, with very solid and concrete bases. It could be market tests, benchmarking competitors or past experiences in similar industries, for example.

■ **How do you distribute the capital? How are decisions made? Is there a "pilot on the plane", or a leader? Who will decide if there is a disagreement?**

Show here that the distribution of roles is clear, and that there should be no possible blockage when making major decisions.

■ **What is your exit strategy and what comparable can you give in the addressed market?**

As a start-up shareholder, you will most often be able to resell your stakes through.

- The buyback of your shares by another company: concretely, most of the capital is sold to a buyer, an established company, interested in the target company. This mode of transfer is the most common in the case of start-ups and allows the joint exit of the shareholders. It can be very advantageous, because the acquirer pays a control premium to the shareholders, which allows them to realize a capital gain.
- The initial public offering (IPO) of the company: this is often the ideal exit, for the company as well as for the investor. The company enters the primary financial market. Its valuation in the market allows shareholders to benefit from new liquidity for their shares, and to realize a capital gain in the event of resale.

In any case, and whatever the questions, here are some tips:

- A project leader who presents himself in front of investors will be able to answer all these questions, provided he is prepared for them and pays attention… The good strategy is to be direct in your answers, to give facts and to seek to reassure (both in the content and in the form of the response).
- Listen openly to remarks and do not let yourself be thrown off balance by questions that can be incisive. This feedback from professionals is the best way to progress and correct your project to make it more solid.
  - Take advantage of this feedback to improve your presentation. The investor might have identified a gap in your model that needs to be fixed. Rather than confront, use this information, and address it.
- If the investor asks you about a point and tells you that you should have included it in your presentation, please do not tell him that you do not find it relevant to indicate it, because if he asks the question, he has clearly indicated that this information is important for him. This happens often, specifically in the case for the question of the amount sought or the valuation. You are there to raise money so do not forget to talk about it,
- Also, do not get drawn into a lengthy one-to-one discussion. Be concise in your answers.

■ When you do not know or still have several leads in mind, do not try to pretend otherwise. You would lose all credibility and irritate the investor by letting him believe that you take him for a fool. On the contrary, turn the situation to your advantage, by showing:
  – Your frankness and your transparency,
  – Your openness to the fact that nothing is completely fixed and that certain strategic decisions will be made later, with the help of the equity investor. Investors know well and say themselves "that a business plan is rarely, if ever, respected". They can understand that some options are still open. Provided of course that the viable options are clear to you, as well as the elements that will intervene in the decision to choose one option rather than another. On the contrary, they can be annoyed if you give the impression of knowing everything,

■ While remaining concise, show that you are passionate and that you love what you do. This aspect is too often neglected by entrepreneurs, even though it is one of the key factors for the success of the project.

Notes

1. Valuation: The valuation is the price of the company on the market at a given moment. This value takes into account the past and the expected future, but also the market value (price that someone is willing to pay …). You must keep in mind that, as long as it is not sold, the price of the company is "virtual."
2. Closing is the last step in a business sale process. It corresponds to the execution of the contract and the actual completion of the sale/acquisition.

Chapter 11

The Art of Negotiating with VCs

Once the VC has confirmed interest in the investment, the negotiation phase begins. Some of the topics which will need to be negotiated with include the valuation, financial and legal, the shareholder's agreement and many more.

Every day we negotiate. It is an essential **soft skill** to maximize our teamwork, create a positive environment and resolve conflicts.

During recent international events, we have seen a reemergence of using a style of hard uncompromising negotiations. The individuals who use this approach to normalize confrontation, rupture and espouse a "take it or leave it" attitude in their negotiation, as opposed to the softer "diplomatic" way. This has been so prevalent that they have succeeded in anchoring this vision in the ambient discourse, inspiring a distrust of the diplomat (a person who knows how to conduct a business with tact), mocking them as a simple wordsmith, a hypocrite never revealing his true goal, endlessly beating around the bush.

This caricature is, of course, as coarse as the characters who espouse this approach and is based on a misunderstanding of the profession of diplomacy; there is much to learn, it goes without saying, from diplomats. Large and small negotiations are an inescapable part of any professional life. In short, be diplomatic and engage your potential investors with honesty and professionalism.

**Philippe weighs in on pitching:** I remember an investment pitch which I left during the pitch as some of the people around the table were very "arrogant". During this first meeting with several co-founders, things were proceeding very well. The business plan was ambitious, but technologies, results and the teams were of excellent (scientific) levels. I was in the presence of a company that planned to develop a therapeutic product in a particularly interesting therapeutic field. After an hour of presentation, a discussion opens on the feasibility for the investment fund to invest in the company. Coming to the topic of the expectations in terms of valuation (opening the capital offered to future shareholders), one of the co-founders stated quite bluntly that below a significant double-digit figure in millions of euros or dollars, there wasn't any need to negotiate further. At that moment, I got up and indicated that we could therefore put an end to all discussion. While the person in question started to backpedal, I was already concentrating on the next opportunity (next deal in the deal flow).

While his valuation might have been defendable, or discussable, his positioning established a context where negotiations were going to be arduous. Companies often over-evaluate their market and IP, and an entrepreneur who engages in pre-negotiation with such a statement has positioned himself quite clumsily. Negotiation is a two-way street.

As per Figure 11.1, to negotiate well, do not necessarily face each other but rather sit side by side. After the negotiation, you will get into the hard part with the application of the business plan and its quirks. Having a real partner on the board is a real asset to achieving success.

11.1 Seven Principles for Effective Negotiations

I've spent years negotiating many types of deals and living with the results as well as observing the negotiating styles and skills of other leaders. With that perspective, I thought I would share what I have learned about being an effective negotiator:

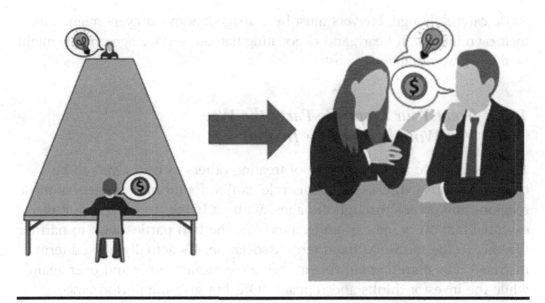

Figure 11.1   Negotiate "side by side" in collaboration versus in opposition mode.

11.1.1 Create a Relationship of Trust between Both Parties

This first principle is an essential condition for the success of a negotiation. However, one must remain attentive to the fact that an agreement should never be concluded solely on trust, omitting the merits of the case under discussion. It is essential to fully understand the ins and outs of the elements to be negotiated. Valuation is one thing, but it is in the clauses of the shareholders' agreement that the details are key: stabilizing the capital, preparing for the exit and protecting investors and founding partners.

11.1.2 Know Well What You Want to Negotiate

Valuation is really a fairly simple element. Despite everything, entry into the capital amounts to "a marriage" and the "common life" must be organized. This is the role of the shareholders' agreement. It is especially important to understand the stakes of the main clauses that exist and their usefulness: stabilize the capital, prepare the exit and protect the investors and the founding partners. The negotiation of the pact requires a good understanding of the consequences of the negotiated clauses. It is therefore important to be accompanied and to be able to benefit from the advice of a lawyer who is used to it.

Be careful though: lawyers must be managed. Some lawyers might have their own interest at heart, and negotiating flat-rate service agreements might be a good way to manage costs.

11.1.3 Treat Your Opposite Party the Way You Would Like to Be Treated

The Golden Rule is the principle of treating others as one wants to be treated. Various expressions of this rule can be found in the tenets of most religions and creeds through the ages. Without being stereotypical, it is essential that everyone can understand that the two parties wish to partner together in the form of a fixed-term association. It's actually a fixed-term marriage. The entrepreneur repeats "give me money" over and over again while the investor thinks and repeats "OK, but give me performance" in return.

11.1.4 Keep in Mind that You Can't Negotiate Everything, Be Strategic!

Don't think you can negotiate everything: the shareholder agreements proposed by professional investors are relatively standard. There is not always of latitude available for making changes. It is easier to negotiate the adjustment of a clause than its deletion. You must also have in mind, in the list of points to renegotiate, those which are critical for you, and those for which you are ready to give up, in order to show your goodwill.

11.1.5 Go beyond Positions to Understand What Drives Them

At the beginning of a negotiation, it is common for each party to take a more extreme position than necessary.

For both, the objective of this method is to keep room to maneuver. As the negotiation progresses, one of the parties agrees to adapt its position in the hope of moving things forward. At that moment, the other party understands that the initial request of his interlocutor was exaggerated, and that he therefore has this leeway.

All in all, it is important to understand that as you are dealing with a professional investor, your specific requests will require validation by the fund's investment committee, and the person in front of you may not have final authority to decide.

11.1.6 Do Not Follow the Proverb "an Eye for an Eye, a Tooth for a Tooth"

Negotiation is not a battle, which means that there is not necessarily a loser and a winner, and you don't have to counterattack or defend yourself. Do not raise your voice, at the risk of the situation escalating.

Make sure you are accompanied when you negotiate, as this insure you have more than a pair of ears listening to the conversation, and gives you the opportunity to discuss and bounce ideas between negotiation sessions. Also, make sure you are able to benefit from the advice of a lawyer who has experience in these types of negotiations and who can act as a filter to avoid negotiating directly if there is a tense situation or topic.

11.1.7 Listen to Understand, Speak to Be Understood!

This is typically the case when discussing about valuation. When talking about valuation, it is important to distinguish the "pre-money value" from the "post-money value" of the company:

- The company has a so-called pre-money value before the contribution of the investor.
- Once this contribution has been made, the new value, known as "post-money", is equal to the pre-money value plus the investor's contribution.

In any case, do not forget that the valuation is the price of the company on the market at a given moment. This value includes the past and the expected future (performance), but also the market value (price that someone is willing to pay…). You must keep in mind that, if it is not sold, the price of the company is "virtual".

Never forget that the objective of the negotiation is to find a compromise to carry out a financial closing (closing the deal) and to have the transfer of the amount corresponding to your "give me money" objective.

11.2 Seven Tips to Convince an Investor

11.2.1 Assess Your Needs before Going into a Negotiation

Raising funds is just like any objective you set in life: if you don't know what you want, you won't achieve anything. Succeeding in correctly assessing your needs is not always easy, and achieving a valuation of the company is even more complicated.

However, before starting any real discussion, it is essential to have a very clear idea of the amount sought and the acceptable dilution. Not only will the dimensioning of the capital requirement orient you quite strongly on the type of investor you are going to talk to, but any backpedaling afterward could prove fatal.

Indeed, if you estimate your need at $2M you will base your business plan on this figure and explain to potential investors why you absolutely need this contribution to succeed. If you realize you've aimed too high and should have made a smaller first fundraising effort with a different strategy, a great balancing act awaits you! It is always complicated to explain that in the end, you will do just as well with $1M as with $2M.

If you are not sure of your need, it is better to plan a drawer strategy, starting for example at $1M and ending at $2M, especially if you managed to generate interest in your company. The dilution will often, in the end, be very close, because in a seed round, the valuation is often deducted from the amount raised and the percentage that the founders are ready to leave to new entrants. It is therefore not uncommon to increase the target capital raised without really affecting the dilution.

All in all, precisely know the amount the company needs and anticipate future financing rounds. Be ambitious but realistic about the amount raised so as not to be too diluted during the first round of financing.

11.2.2 Never Lie

When you pitch your project, you can sometimes be tempted to embellish the situation, to inflate the short-term figures and to present potential contracts as finalized deals. While it is possible to be overly positive on future revenues for year five of your business plan, it is foolish to present something as finalized and confirmed when it is not.

An investor will generally never forget what was said to him. This is the reason why you see them taking lots of notes during meetings. Raising funds is a long journey. Between the moment an agreement in principle is reached and the moment the money is in the bank account, it often takes several months.

If you announced fantastic short-term figures to make investors excited about your opportunity, and during the preliminary discussions or due diligence the short-term revenues turn out well below the forecasts, you risk losing all credibility in the negotiation.

The best strategy is to keep cards up your sleeve.

Be conservative on your short-term business plan, don't talk about all the major contract you're about to sign, don't mention an order from a strategic client that you are unsure of. If you miss everything, at least you won't have to explain yourself. But if, on the contrary, everything goes as you planned, then you will energize your fundraising process with an uninterrupted flow of good news. And when we go from good surprises to good surprises, we want to be on your side!

Always do what you say you are going to do, be straight in your boots. You must be a model of integrity. Would you like to entrust your money to someone for whom reality is negotiable?

Finally, never lie to an investor on a so-called term sheet sent by another VC. While it's important to show that you're being courted, announcing that a term sheet is in your possession when you don't is a recipe for a future disaster. Everything always ends up being known, and if you use this subterfuge to speed up the signing of a deal, the trust with your investor will be broken and he may be less inclined to help you the day you need his help. As Jean-Paul Sartre said, "Confidence is gained in drops and lost in liters".

JF weighs in on negotiating multiple deals: On the other hand, the ideal negotiation situation for an entrepreneur is to have multiple term sheets developing at the same time. Having only developed a single lead can lead you to a situation where you only have one option, which is "take it or leave it". Being able to manage multiple conversation, while also developing a common timeline with all the funds involved (slowing down the faster player, and accelerating the slow ones), will definitely increase your leverage. But, coming back to Philippe's point: don't invent non-existing term sheets, as the VC can just as well call your bluff, leaving you with nothing.

11.2.3 Be Authentic and Know How to Listen

It is useless to play a role, to posture or to invent a personality. What an investor will evaluate first, especially for a project in the development phase, is you, your cofounders and your team. If you give a distorted image of reality, because you're looking for inspiration from other speakers or because you've watched too many great speakers and feet insecure, you'll only get trouble and confuse your audience.

Being authentic also means being vulnerable and showing your weaknesses.

No sane investor expects to see an invincible machine. Talking about the things you can't do is as important as highlighting what you can do. Being aware of its limits is a quality that will be appreciated by investors.

There is indeed nothing worse than an entrepreneur who does not know how to ask for help, and who ultimately does not know how to surround himself with additional expertise.

When raising funds, you find yourself talking to dozens of investors, all of whom have a different approach and understanding of your project. If you're not able to listen to what you're told, not only will you be missing out on great advice, but you'll also be showing investors that you're not ready yet.

11.2.4 Know When to Walk Away

You should never enter a negotiation with a position where you cannot walk away. If you do, you are giving up your leverage. As such, knowing how to say "No" to the demands of investors is an important element to have in your mindset. Accepting any conditions at all costs can sometimes

JF weighs in on the evolution of your pitch: As you do your pitch to different investors, it will evolve. A pitch is not static, and all the comments you will receive during the negotiations and pitches will lead you to constantly rework your presentation.

As such, take note of the comments and questions you get during your pitch sessions. It can also be a good idea to bring a partner to pitch meetings so they focus on reactions from listeners, and then help you optimize your presentation.

lead to too much risk-taking. It is therefore essential to surround yourself with the right people and to agree to work with legal and accounting professionals in order to secure the operation and anticipate the scope of the commitments made.

11.2.5 Stay Focused

As you build and grow your business, many opportunities can arise. You can easily find yourself in a mindset where you reflect "What if we developed this application" or "What if we opened this vertical market tomorrow" or "What if we launched a new generation of products". While it is always beneficial for an entrepreneur to remain open and attentive to what is happening, when it comes the time to convince an investor, it is much more efficient to concentrate rather than being everywhere.

An investor will already struggle to understand what you plan to do, analyze the market, and model the different revenue assumptions. If on top of that, you keep gossiping about all the amazing and wonderful things you *could* be doing, you will be perceived as unready for VC capital.

You need to be very clear about the value proposition and as simple as possible about your execution strategy.

Finally, it is essential that your roadmap and the use of proceeds are justified based on solid estimates. The best life science entrepreneurs from our point of view are those who have remained focused on their roadmap and who have been able to manage cash as well as possible through rigorous anticipation.

11.2.6 Be Resilient

A good fundraising process will take several dozen pitches. It's several dozen opportunities to discuss your project constructively, but it's also (and above all) dozens of opportunities to get a multitude of negative feedback or "no". And at the same time, you must manage your business, support the teams and put out the various daily fires.

Three ways to improve your resilience (and confidence) when it comes to fundraising:

- Practice, practice, practice,
- Get okay with "no",
- Don't be afraid to ask for support or help.

Attend networking events within the local startup, entrepreneur and investing community. Mention your startup or existing business naturally and let a conversation develop organically.

11.2.7 Find a Lead

A lead investor in the context of fundraising is the investor (or more often the investment fund) who will take the lead in negotiating the terms of the investment, the clauses of the shareholder, and the various due diligences. It is generally the one who will invest the largest investment ticket.

It all depends on the overall context of private equity at the time of your fundraising and on your geographic location. There are cycles (crisis of euphoria then crisis of wait-and-see attitude). But without a lead, there is no possible investment. Finding a lead is therefore a crucial step in the fundraising process.

A good strategy for finding a lead is to convince a successful entrepreneur with great legitimacy in your field to join your team, in an advisory capacity, for example. His presence will reassure the potential lead and help him go part of the way. This "star" entrepreneur will bring a strong added value in addition to his financial contribution, so it will often be relevant to offer him a place on the board.

Remember that an early-stage investor, and particularly a follower, will be more likely to invest if a lead or co-lead is physically present. As a seed investor, it is essential to be close to the management teams. A regional or national leader is a prerequisite for convincing other national or international investors.

11.3 Some Key Issues that Will Require Negotiations

You will negotiate many elements as you get closer to the final closing, but three of them merit specific consideration: your valuation, the decision-making process control and cofounders/founder obligations.

11.3.1 The Challenge of Valuing Your Business

Successful fundraising is above all about promoting your business. Valuation is one of the key elements in negotiations with investors. While valuation is always a tricky subject, poor valuation can have significant repercussions for founders. The valuation can (and will) be negotiated with the investor. In some cases, investors will insist on adding correction and adjustment mechanisms in order to protect themselves against a fall in valuation (see Section 10.5.1 for more details on typical clauses that are usually negotiated). During this negotiation, it is necessary to keep in mind that several other investment rounds are possible in the future, and to consider both the current and future negotiation rounds when building your valuation.

When starting negotiations around valuation and investment amounts, a good approach is to take a metric such as the percentage of the capital (% of shares) that you wish to open to new shareholders versus displaying a face "pre-money" value. This could allow for example to negotiate on the size of the investment ticket rather than negotiating on the valuation. A good way to leave the door open to future negotiations is to specify a range such as 20% to 30% of the company's capital for a first round.

Early-stage investors tend to use equity warrants, convertible loans and common shares for seed or first rounds of investing. Convertible loans are a financing method that is becoming increasingly popular for start-ups: startups receive a loan that will be converted into equity (common stock or equity shares) at the next qualified financing round. Investors who provide convertible loans can receive interest payments and a discount on the stock price they pay compared to the price incoming investors must pay in the financing round. Convertible loans often come with a cap—a maximum valuation at which the loan can be turned into equity. There are two key terms that are generally the provisions in a convertible note or convertible equity instrument that are the most significantly debated at the time of the deal: the cap and the discount. In addition, there are several other important terms in convertible notes and convertible equity, the cap (the cap is the maximum valuation at which the shares will convert to equity), the discount (the discount rewards the convertible note holder for taking a risk on an early-stage company by allowing the holder of the note to convert at a lower price than the investors pay at the conversion round). To get a better understanding of what can be negotiated, we have included a summary of negotiation points (Table 11.1).

You might notice that you can simplify issues relative to the valuation of your company at the time of financing by projecting the conversion on a future value of the next round of the company. Instead, with this approach, you will find yourself negotiating on other important elements such as discount and interest rate.

Table 11.1 Typical Terms Subject to Negotiation in Convertible Notes and Equity

Terms Applicable Only to Convertible Notes	*Terms Applicable to Both Convertible Notes and Convertible Equity*
• **Time to Maturity:** the duration of the debt instrument (i.e. when the note holder can "call the note" and demand repayment) – this is typically between 1-2 years; • **Interest Rate:** the interest rate attached to the debt that will accrue over time (and will be incorporated into the conversion) – this ranges significantly between 2 or 3% and can reach upwards of 10%; • **Ability to Pre-Pay:** the startup may or may not be permitted to pre-pay the loan back to the investor instead of allowing the investor to convert the loan into equity.	• **Rights Upon Acquisition:** usually the investor will be paid back extra upon an acquisition of the startup by another company. • **Conversion Event:** the definition of the round at which the conversion occurs (this is usually described as a minimum aggregate investment); • **Warrants or Other Future Rights:** to entice investors to take a risk on an early-stage startup, the investor will sometimes receive warrants that allow the investor to purchase additional stock at a future date.

11.3.2 The Issue of Control

Fundraising always results in a distribution of shares to one or more investors. This new distribution will logically have the effect of diluting the participation of the founders as well as investors already present in the capital of the target company of the investment. As dilution is an inevitable consequence, it is essential to be aware of the resulting shifts in decision-making power that such dilution entails for all the partners of the company already present in its capital.

Although investors generally do not seek to take control of a company, it is normal to impose prior consultation mechanisms or veto mechanisms on the founders before taking certain strategic decisions. Distribution

Jf weighs in on development costs: We didn't touch on this earlier, and might disappoint some entrepreneurs, but here it goes: you should never negotiate based on development costs. These are sunk cost, and you will see their value reflected on the value they created on your company. So, if you spend $1M, which generated two patents and a device ready to commercialize, you negotiate on the value of the patents and the device, not on the $1M, used to develop these assets. This is especially hard when dealing in the therapeutics space, where you might have millions of dollars, with little to show for it, but it is a hard reality of valuation.

of dividends, fixing of executive compensation, recruitment of new key employees, investments made by the company, loans to be made are all topics investors will want to have their say in. How these topics will be handled will generally be subject of negotiations.

All these elements will be found in the shareholders' agreement which will bind the founders to the investors; it is therefore essential to negotiate it well and to only commit once all the reciprocal obligations have been fully understood.

11.3.3 Founders' Obligations after an Investment

For investors, the expertise of the founders is often a key factor in the success of the project. This is why investors will require many commitments from them. This is particularly the case for non-competition and exclusivity commitments. Through this type of clause, investors want to be certain that the operational partners of the company will put all their energy into the success of the business project, and not divest themselves in parallel projects.

When negotiating the terms of the investment, it is also customary to add clauses which will prevent the founders from selling their shares or, on the contrary, will force them to sell them. Some investors want to see an inalienability clause in the pact; the latter prevents certain designated partners from selling their shares for a given period or until the occurrence of a particular event.

In other cases, it is necessary to provide mechanisms anticipating the exit of the founders from the company. These mechanisms are generally contained in the clauses relating to the leaver. Thus, various hypotheses such as resignation, permanent incapacity, dismissal for serious misconduct or breach of a non-competition or exclusivity commitment may be causes for the forced transfer of shares.

11.4 Closing Words

Bringing an investment fund to invest its capital is not easy. Like any change in shareholding, this step must be negotiated carefully. Post-crisis, financiers have adopted a more wait-and-see and risk-averse attitude.

Raising venture capital funds has become an art, even the most skilled negotiators can find it difficult to navigate. Careful management of power dynamics is crucial to getting the terms you want and need. The future of your business should be based on careful negotiation of venture capital with the goal of a fair deal and hopefully a new partnership. In the business world, the journey is long, and so are the relationships. Stay authentic, be yourself, and foster sincerity in your connections.

The advice presented here does not in any way claim to be exhaustive, nor to provide answers to all situations. In the end, only your instinct counts. As indicated previously, entrepreneurship is Darwinian, Schumpeterian and Brownian, but it is up to you to find the right leverage effects to create, seize opportunities and direct evolution towards a "yes"!

Chapter 12

Eight Classic Mistakes Life Science Entrepreneurs Make

Business creation is at the same time Darwinian,[1] Brownian[2] and Schumpeterian.[3] Of course, you have to work a lot, but you really have to be lucky! Entrepreneurship is complicated, risky, and requires luck, resilience and adaptation.

The original plan is rarely the one that will unfold throughout the life of the business (Figure 12.1). It is always difficult to anticipate the unpredictable but walking the learning curve is also a question of preparation. Be ready for the unpredictable …

We have had the opportunity to examine, review and interact with hundreds of start-ups through the years. We have seen them rise, and fall, and after a while, have started developing a quick checklist of the basic mistakes that entrepreneurs make. To avoid these mistakes, it is good to know them. As Benjamin Franklin once said, "There are three things extremely hard: steel, a diamond, and to know one's self". Some of these mistakes include:

- Not the right management team
- Not enough funding or lack of fundraising strategy
- Cash flow problems or poor cash management
- Flawed business model
- Regulatory or legal issues
- No market needs or urgency for a startup solution
- Lack of intellectual property or patent protection

DOI: 10.4324/9781003381976-12

Figure 12.1 Your plan versus reality.

- Expansion too fast
- Ineffective marketing/business development

We've merged these elements into eight mistakes, which we will share in the next few pages.

12.1 They Underestimate (or Overestimate) Their Pitch Deck

This is a very common scenario; some entrepreneurs do not want to spend too much time on their pitch deck because they prefer to stay focused on their business. Staying focused is a noble endeavor, but the pitch deck should not be overlooked since it is one of the first pieces of information that an investor will receive about your company and as you know, the first impression weighs heavily in the rest of the relationship.

Other entrepreneurs spend an inordinate amount of time on this document and heavily rely on it. The Pitch Deck is only one element of your toolbox and one must make sure to prepare other documents and show your ability to execute. Few entrepreneurs raise funds with their Pitch Deck alone.

☞ **Key Takeaway:** The Pitch Deck reflects your startup, and in particular of the development and dynamism of the commercial activity. Spend the time you need on it, and ensure a great first impression when fundraising.

12.2 They Believe that Narrative Is Not Everything

Storytelling is one of the most powerful tools that entrepreneurs have in their toolbox. A well-crafted story can inspire investors in wanting to learn more about your startup. These stories can also help you stand out in a crowded market, especially after a long pitch, where the investor has had to listen to dozens of presentations that day!

Many pitches that we've listened to are focused on numbers, analytics, ratios, scientific data, return on investment but completely forgo storytelling. They come out as dry, boring talks, and give yet one more excuse for your listeners to turn to their smartphone, losing interest in your pitch. Yet, stories are how we connect with each other and how we build relationships. When used effectively, storytelling can be a powerful tool for startups to connect with investors, and make sure they want to talk with you again!

So, you might ask yourself, when doing scientific research or developing a deep tech product or service, why is it so important to work on your story?

 **Jean-François weighs in on the three What's of storytelling:** We mentioned this in Chapter 8, but it bears mentioning again: Presenting your company to an investor is about taking them through the **three what's**: 1) what happened (the problem), 2) so what (why is this important) and 3) what now (how are we solving this). Following this narrative creates an expectation, kind of a "What happens next?" feeling we get when we watch a movie. Build up your hero (your team, your technology, your IP), and show how they can vanquish the villain (your competition, your technology challenge, the problem you are trying to solve).

☞ **Key Takeaway:** *Your story is the key support for the four pillars of your startup journey (as seen in Figure 12.2): the two external pillars, which are your customers (design, products, partnership, marketing, sales, etc.) and your capital (VC's, investor relations, your strategic committee), as well as the two internal pillars, which are your employees (internal communication, operations, alignment with shareholders, etc.) and talents (hiring, recruitment strategy, etc.). Your story will be key in sharing your vision to these four pillars.*

12.3 They Believe There Is No Competition to Their Innovation

We always have competitors. You should not underestimate your competitors.

If it is not direct, it will be substitution. If it is not a substitution, it will be indirect. Such is life. Let's take a concrete example. When Henry Ford invented the car, he had no competitor, strictly speaking, but what about horses and carriages? Your competitors aren't always who you think they are (Figure 12.3).

In France, what's more, when you're at the head of a start-up, it's very easy to become arrogant and dismissive towards potential competitors "who have ridiculous technology", because we don't watch enough what happens next. Worse, when an entrepreneur learns of them, they don't always take them seriously, because "it's impossible for his product to be as good as mine". What if I told you that a product that is less efficient than yours can hurt you a lot if the vision and the associated communication are efficient?

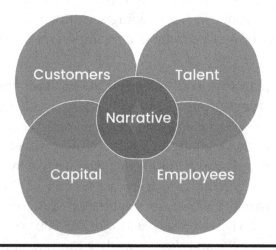

Figure 12.2 The four pillars of your startup.

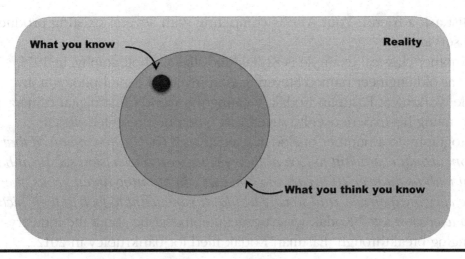

Figure 12.3   Do not underestimate your competitors.

 **Philippe weighs in on misevaluating your market:** In the 2000s, a brilliant French engineer developed a revolutionary product: an mp3 player. Indeed, I had bought a Jukebox player in the 2000s (the Jukebox 6000's menus were complex, and its firmware buggy)! It was technically incredible except that for ergonomics, design and user experience, the engineers in question clearly thought that it was the fifth wheel of the carriage! The French company Archos did indeed invent the "portable media player" but it arrived too late... A few months before its commercialization, Apple presents the iPod.[4] The mistake of Archos engineers: not necessarily caring about interactions and ease of use of their products, all things that have made Apple successful, since that seems to be what matters most to the greatest number of customers. In life science, how often do people developing an innovation forget what matters most to the user (the patient, the client, the user)?

One way to frame the question is that you can have a competitor by technology (does the same thing with a similar technology) or by purpose (accomplishes the same goal). For example, if you invented an APP that scans your hair to diagnose hair loss. Your review of the Appstore has shown that no other APP does the same thing, but you are not the first product to do this function: a client could decide to see a doctor, self-diagnose or just ask a friend. Your APP is competing with several existing products and services.

Another classical example is Kodak and digital photography. In 1975, a 24-year-old engineer named Steven Sasson invented digital photography while working at Eastman Kodak, creating the world's first digital camera. Recounting his experience, he stated that when he presented digital photography to a number of Kodak executives: *"They were convinced that no one would ever want to look at their pictures on a television set. Print had been with us for over 100 years, no one was complaining about prints, they were very inexpensive, and so why would anyone want to look at their picture on a television set?"*[5] Kodak wasn't exactly enthusiastic about the industry-changing breakthrough. Eastman Kodak filed for bankruptcy in 2012...

☞ **Key Takeaway:** *There is always competition. It can be a technology or it can be by purpose. A correct appraisal takes into account both types of competitors.*

12.4 They Raise Funds Only When They Need Them

Less experienced entrepreneurs raise funds only when they need them. If a company tries to raise funds because it is about to run out of money, but has not reached a value creation milestone, this negatively impacts the interest of investors as well as the business development of the company. As we have seen, fundraising is a long process. Also, it is essential to do it sufficiently upstream: when you are fundraising, you create a news flow throughout the fundraising process to maintain the interest shown by potential investors. Many entrepreneurs instinctively focus on fundraising primarily when cash flow issues are noticeable.

It is crucial to objectively consider the time it will take to reach value creation milestones. Financing milestones must correspond to elements where you reach key development stages. In essence, this means identifying the amount of funds needed to achieve each stage of value creation.

As indicated in Chapter 10 – "The Venture Capitalist's Rulebook to Investing", it takes a lot of time to establish relationships, to present your story, to allow investors to carry out due diligence, to negotiate terms, to prepare legal documents, to close the deal and to transfer of funds to the company's

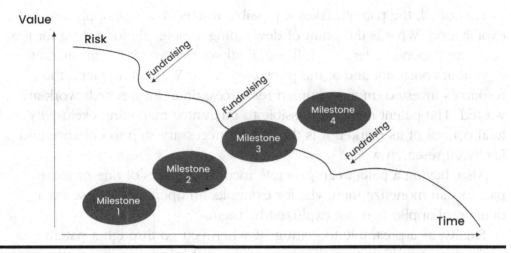

Figure 12.4 Plan your fundraising after technical or commercial milestones.

account. Also, take into account that if you are rushing through fundraising, you are essentially giving up your bargaining position, as you will most likely be in a position where you cannot refuse an offer. (See Figure 12.4.)

A good rule of thumb, many investors internally apply the "doubling rule". If you believe it will take you 24 months to grow from a Technology Readiness Level (TRL) of three to TRL five,[6] well they consider that it will take 48 months! If the company needs 1M€ for its first prototype, most investors consider that it will need at least 2M€.

☞ **Key Takeaway:** *Start raising capital at least 12-24 months before you need the capital, depending on the size of the financing round. Concretely, you must always be in "fundraising" mode, whether you need it or not, because you can always say "no" to an offer but saying "yes" implies having offers on the table.*

12.5 They Underestimate (or Overestimate) the Protection and Value of Intellectual Property (IP)

Without a patent, a business cannot effectively raise venture capital in Life Sciences! To find investors, you must tell a convincing story and reassure them. This is where IP comes in. Life sciences is a high-risk investment, so investors will take a closer look at companies that own patents.

First of all, the patent makes it possible to ensure a monopoly of exploitation. What is the point of developing a molecule for a drug for ten years, and spending tens of millions, if all your competitors can market it without constraint and at the price they want? Without a patent, the resources invested (money, human resources, time) for research work are wasted. The patent makes it possible to guarantee marketing exclusivity and total control of its market. It is therefore a necessary step to enhance and justify all research work.

Also, having a patent can generate income. Holders of one or more patents can monetize them via, for example, an operating license on a domain of application not exploited by them.

Finally, as a preamble to getting IP, when you go through a patent application, you have to make sure you are not infringing on another patent. As such, it helps establish that you are not developing a technology that overlaps a competitor's IP space.

In summary, patents are protective barriers for a sustainable business and are an intangible financial asset.

You should view patents and intellectual property as strategic business investments, not costs. Startups are often interested in the industrial property strategy as a second step. However, filing patents in the seed phase should be a priority. Innovation is boiling, the players are multiplying, the work is accelerating and the rule remains the same: the first to file a patent, and what is more robust, will obtain a monopoly of exploitation.

On the other hand, many entrepreneurs will overvalue their patents. A cash-strapped company with a portfolio of patents will be ill-positioned to defend them. A more common occurrence is a company securing a patent in a space with huge upswing, and then basing their valuation on that single patent. More than once, we have met entrepreneurs whose $100M valuation of their two-man start-up was based on a patent in the oncology space... a patent is a first step in developing your company, and if you cannot demonstrate your ability to use or defend this patent, its value will be severely compromised.

☞ **Key Takeaway:** *The IP strategy is an integral part of the evaluation process for investors. It has as much value as defining the size of the target market or demonstrating a partnership with a reference laboratory which legitimizes the research.*

12.6 They Overestimate the Value of Their PhD or Business Experience

This is often referred to as the omniscience syndrome.

Founders are often scientists with expertise in their field. Their technical background makes them suitable for working on the development of the company's technology or product. However, that might not make them the best candidate for a CEO role. Sometimes a founder doesn't have the business experience to fulfill this role effectively.

If you don't have the experience and expertise to run the company, part of your job as a founder or co-founder is to find a capable CEO you can trust to implement the company vision. Identify what you need in a CEO and put in place a way to vet candidates. Ask nothing less than 100% of them, because anything less than that can lead to failure.

Additionally, it is crucial that you do your best to hire a strong management team or establish a division of labor/roles between you and your co-founders. Who will assume which roles? Are there gaps in your knowledge and expertise and those of your co-founders that you can fill by hiring someone else?

Whatever you do, do your best to support the right people because the right people can help you weather the storms you will face as you try to grow your business. Generally, the most important areas of expertise are finance, business development, marketing, regulatory and industrialization.

☞ **Key takeaway:** *Many life science company failures are not primarily technology-related but primarily due to management issues generated by a weak management team. Being able to identify our limits, and being able to recruit good leaders is an important task in the company.*

12.7 They Neglect Cash Flow Management

Ensuring you have a good cash flow management means having real-time view of your figures, and more precisely on the expenses incurred and the resources available. The objective is to anticipate as much as possible in order to ensure the solvency and profitability of the company in the short and medium term.

For a business manager, it is essential to have a permanent view of the cash position. If creating a business plan makes it possible to establish a development strategy, it is also imperative to be able to regularly compare forecasts with reality.

The annual budget forecasts the evolution of the company's cash flow for the following year, month by month. Most often compiled at the end of the financial year on the basis of the financial years already carried out, it groups together the inflows and outflows of money planned for the next financial year. These are, on the one hand, forecasts of receipts (sales, new financing, etc.), and on the other hand, disbursements (investments, purchases, overheads, salaries, etc.).

The provisional cash flow plan is a table that must be updated as the financial year progresses, on a monthly or weekly basis. Its use allows dynamic management control to adjust the means implemented. In other words, the company needs to have a monthly cash flow forecast, over 6 to 12 months. The comparison of these forecasts with the actual figures will make it possible to better anticipate receipts and disbursements, and to update the plan according to the evolution of the company's activity.

Cash management was the leader's lethal weapon in post-COVID-19 2022 and 2023... Cash management is a crucial element in the life of the manager since it ensures the financial balance of his company. It should never be done superficially.

Two ratios that are often used to quickly ascertain if you have good cash flow management are your burn rate and cash runway. The burn rate is the speed at which you are spending your funds each month, while the cash runway is the rate at which you will burn through current cash reserves at the current burn rate. So, for example, if you have 1 million dollars in cash, and you have a burn rate of 200K per month, you have a cash runway of 5 months before you fold your company. Knowing your burn rate and cash runway at the top of your head means that you know how much cash you have, and how fast you are burning through it.

☞ **Key takeaway:** *Manage your cash flow carefully, and calculate your burn rate and cash runway so you can quickly illustrate that you have some level of control over these aspects of your business.*

12.8 They Neglect Their Brand

It's very surprising (and discouraging) to see how little effort companies put into their own branding. Consistently, we'll work with entrepreneurs who develop complex technologies, are engaged in advanced research, spending countless hours building their pitch deck and practicing it, mapping out their five-year plan in every complex detail, and then slap on the first name that comes into their mind on their product and calling it a day. In the haste to reach market, or meet investors, branding is often neglected.

Abbreviating your name and adding "biotech", "genomic" or an "AI" to the first idea that you came up with is not a branding strategy.

While branding might seem superficial, it is often the first point of contact a partner or investor has with your company. You wouldn't go in a meeting wearing your worn jeans and your old T-shirt, so why would you send a slide deck with a botched logo and a company name like "JFPT Biotech.AI"?

This is doubly true in the digital app space, where a quick search in the Play Store can bring up pages and pages of similar and competing apps. While a consumer might take an in-depth look at the app, the first selection criteria in an APP store is one with a strong brand name.

 **Jf weights in on branding:** Some tips to improve your brand include focus on brevity (shorter brand names are easier to write, read and remember), meaning (make sure the purpose of your company and product is easily understood) and likeability (find a brand that elicits positive feelings). Then, take it out for a test drive, and test it with colleagues and friends, then with people unrelated to your company. Make sure people understand what you are trying to convey in your message.

☞ **Key takeaway:** *While a good brand won't save a bad company, it is nonetheless an asset to your company, not a hindrance. As you take time to develop your product, take time to develop your brand to make sure it is brief, meaningful and likable.*

12.9 Key Learnings on Key Takeaways

A startup must talk about its project to convince investors, employees or business partners. If a good physical and mental preparation is necessary, the realization of a relevant pitch deck is just as important to offer an attractive presentation of your project, your team and your company to your audience.

The decision to make the investment lies in the hands of the investors, but what you can do is create an impressive startup pitch. The manner in which you present your product in a pitch deck is extremely important. When an idea is pitched to an investor, the data, figures, and numbers could be forgotten but a great story (storytelling or narrative) could have a lasting impact.

Don't underestimate your competition. A common mistake that most start-ups make when researching their competition is overlooking substitutes (product or service that can be easily replaced with another by consumers or users). Often entrepreneurs or technologies focus only on other companies that have very similar business models.

Seed investors know that most life science company failures aren't primarily about technology, but mostly about management issues with a weak management team.

Being able to identify our limitations, and being able to recruit good managers is an important task in business.

Successful fundraising is not a long calm river. This is even more true when this is the first fundraiser. It's a marathon of at least 12–24 months. To maximize your chances of convincing investors, remember two things: preparation and promoting traction (business traction).

Notes

1. Darwinism is a theory of biological evolution developed by the English naturalist Charles Darwin (1809–1882).
2. Brownian motion is the random motion of a particle as a result of collisions with surrounding gaseous molecules (Robert Brown 1773–1858).
3. Joseph Schumpeter (1883–1950), who defined competition as a dynamic process wherein firms strive to survive under an evolving set of rules that constantly produces winners and losers.
4. Nelzin-Santos, Anthony. Archos, les limites du « made in France », 14th of February 2012, https://www.macg.co/unes/voir/130972/archos-les-limites-du-made-in-france? (Last verified 14th of June 2023)

5. Zhang, Michael. What Kodak Said About Digital Photography in 1975, September 2017, https://petapixel.com/2017/09/21/kodak-said-digital-photography-1975/ (Last verified 14th of June 2023)
6. Technology readiness levels (TRLs) are a method for estimating the maturity of technologies during the acquisition phase of a program. TRLs enable consistent and uniform discussions of technical maturity across different types of technology. TRL is determined during a technology readiness assessment (TRA) that examines program concepts, technology requirements and demonstrated technology capabilities. TRLs are based on a scale from 1 to 9 with 9 being the most mature technology.

Chapter 13

How the World Changed Following COVID-19

 COVID-19 changed many processes and assumptions that we took for granted. In our sector, it impacted both the entrepreneurial process as well as the investment process. As thing around the world slowly settles in, we can start to see a new world shaping. There is clearly a pre-COVID and a post-COVID world when it comes to the start-up space. As a reminder, we have proposed throughout this book that the entrepreneur's journey is a five-step process (Figure 13.1):

Figure 13.1   The entrepreneur's journey.

For each of these steps, we have identified some major shifts which impact both entrepreneurs and investors. Let us take a moment to review these and see how this could have changed some of the more traditional processes and assumptions.

      DOI: 10.4324/9781003381976-13

13.1 How Has the Discovery Stage Shifted?

One the main changes we have noticed in the discovery stage is the notion of speed. We used to live in a world where ideas spent a little more time in the creativity and exploration phase. Today, even the most basic brainstormed idea is rapidly evaluated and subject to quicker scrutiny. The creativity phase has been shortened, and the attention on clearly positioning the innovation on the problem/solution paradigm is emphasized.

As we live in an increasingly remote world, we have also noticed an emphasis on ideas that are suited or developed with distance in mind: These technologies increasingly emphasize both asynchronous and synchronous communications, and their objective is to both accelerate caretaking and save time. From remote-patient monitoring solutions to video-enabled healthcare visits to consolidated patient files, more solutions are being developed with a remote perspective in mind and with virtual company models to support these initiatives.

Investments have followed this trend, with US-based digital health startups raising almost $30 billion in 2021, almost doubling the total investment the year prior.[1]

13.2 How Has the Evaluation Shifted?

When we walked through on how an innovation is evaluated, we defined three main axes: Technology, market and intellectual property. As we reflect on the evaluation process, we can safely say that this step has also accelerated, and short-term perspectives currently dominate.

On the technology front, a technology with a sustainable competitive advantage used to be a sure way to entrench a market opportunity. As we saw during COVID, technologies can be developed at an accelerated pace, and building a company around a great competitive advantage might be viable, but single-advantage companies get less attention than they used to. In fact, products should be built with multiple advantages in mind, and scenario-based discussions where a company manages the risk of one of their technology advantages to another competing technology must be included in the conversations. Since post-COVID, investors from across the investment spectrum look to attenuate risk by focusing on solutions that are not "one-trick" ponies.

On the marketing front, market feasibilities studies and return on investments models now favor shorter revenue cycles. Revenues are no longer planned around a three-to-five-year hockey stick revenue curve, but we are seeing more of a push towards a 12 month perspective. Most investors now look at projects and target those that generate sales within 12 months.

 **Jf weights in on financial modeling:** Thinking back to pre-COVID days, we used to build financial models with a five to seven years perspective. In the renewable energy/energy efficiency space, these models could even be pushed to ten year financial modes. Looking back further, during my university days, I was taught to do 20 years models for revenues and discounted cash-flow valuations. I cannot imagine walking into a VCs office today with a 20 year forward financial statement. Today, a model that goes over four-year needs need to be justified. Products that have long regulatorily approval pathway (such as biologicals), for example, are usually exempted from short term models, but other companies should be able to demonstrate (and justify) some form of revenues in the first few years.

In the IP space, increasing litigation around patents has meant that investors increasingly rely on third party evaluation to get a best perspective on the actual value of the patent. Having a patent is one thing, having the ability to use it is another. The article by Gaviria and Kilic, "A network analysis of COVID-19 mRNA vaccine patents" is a great illustration of the difficult space around COVID-19 patent space.[2] In this article, we can see immediately the interactions between players as patents intertwine in a symphony of licences, collaborations and lawsuits. Hence, having a patent in the COVID-19 space, without understanding the current IP landscape, could lead to potential overlaps or lawsuits in the future.

13.3 How Has COVID-19 Shifted Strategy?

Speed and remote work have scattered talent, resources and clients across the globe. As such, while some businesses still lend themselves to very local models, more companies build models with distributed workforces and clients in mind.

Hence, companies are hiring more talent abroad, use remote technology solutions and work with distributed partners with a lot less resistance than before: Talent and cost are becoming more important when selecting resources, rather than their local availability. You might choose a remote AI expert with a specific expertise to your disease field, rather than "settling" for a local resource whose only current advantage is that he can walk in the office. That same resource, with his specific experiences, might be best suited for another company across the globe looking his expertise. As such, there has been a shift from classical nine-to-five model with offices, to hybrid models that include several remote resources. Of course, distributed teams created additional challenges around communications and sharing data which should not be discounted when implemented.

Furthermore, we used to think of potential markets in a geographic framework: A specific geographic area market would be targeted since it was the closest to the company. The entrepreneur knew the market firsthand, and since he was closest to it, it was easier to physically cater to it. But today, with more remote solutions, both in terms of customer acquisitions and service, we have noticed a shift where markets will often be targeted and assessed for reasons such as regulatory framework, IP Landscape and opportunity to service. Geographic boundaries still exist, but regulatory and IP boundaries are starting to play an increasing role in evaluating opportunity.

Jf weights in on commercialization strategy: As a recent example, I was working with a company developing platforms and products using gold and silver nanoparticles. As is typical, it was targeting its local market as a primary market for its technology. But it then decided to target secondary markets such as Brazil, Spain and Mexico, rather than the "traditional" secondary markets that Quebec Canadian companies usually target (such as the US and France). They explained that while the larger US market was interesting, Brazil, Spain and Mexico already had a tradition and culture of using silver nanoparticles, making it more attractive since market penetration was potentially faster. Also, for some of these markets, Canadian regulatory approval was often enough to distribute and market these products. As such, rather than focusing on traditional proximity partners, market decisions focused on regulatory aspects as well as market trends.

13.4 How COVID-19 Has Shifted the Pitch!

Significant changes have occurred in this space, but differently than what experts initially expected.

Many expected that virtual conferences would emerge and replace our face-to-face interactions. Higher customization in matchmaking, lower costs and participation flexibility all seemed like reasonable arguments. After three years of COVID, what we found is that these tools were not ready to replace live events just yet. Lack of attention during meetings (is that person really listening to me or is he doing emails?), low commitment (*I have had events where less than 50% of scheduled meetings showed up – Jf*) and lack of warmth in connections all encouraged us to rush back to live events.

As such, what we have found today is that we must develop a balance between the two mediums. We are not headed to an all-virtual meeting world, nor are we heading back exclusively to the face-to-face conferences and meetings but rather both in moderation.

The increase reliance has now led to start-ups doubling up on preparations. Now, start-ups must start building their presentation pitch with both the virtual and face-to-face world in mind. Hence, we have seen that many investors now do the first evaluation meeting virtually and move on to face-to-face once they have validated interest and potential opportunity. Remember, from Philippe's earlier explication, VCs are machines that are geared to say no. Hence, being able to quickly review and discard a high volume of potential investments is useful. As such, it is not rare for startups to practice and prepare themselves for both types of pitches.

This has also shifted VC evaluations, and they find themselves increasingly scattered across the map. VCs that once had portfolios restricted to a small geographic area, are now looking at deals based anywhere from Montreal to Paris. Hence it is common for a VC to have a broader portfolio than before. But for some, this is a short-term trend, and VCs are expected to return to their roots and develop regional portfolios rather than developing global ones. While it might be easier than before to look at deals across the globe, VCs often invest their own time and supply their expertise to startups they work with, and while there is some general learning that can be shared, regional expertise is often one of the necessary skills when monitoring a new investment.

Evaluation velocity is another point of contention. Some articles mention that investment velocity has increased, with investors taking more risk more rapidly. But this might be a trend that was apparent during the COVID-19 period itself, rather than a way COVID-19 has shifted the investment cycles. As we slowly move on from COVID, many investors are now increasing their due diligence and slowing down their investment cycle.

 **Philippe weights in on investment cycles:** As I mentioned earlier, slower investment cycles enable us to closely monitor and validate some of the milestones as well as financials and revenue estimates you submitted during your evaluation phase: If you presented an estimated sales traction which would generate $1M in revenues after six months, and you haven't generated a single sale during the whole year of due diligence, it **will impact** the investment decision.

13.5 How Has COVID-19 Shifted the Growth Cycle?

From an entrepreneur's perspective, if there is one thing that has shifted post-COVID is how rapidly healthcare systems evaluate, test and integrate technologies. In the past, a project would have to go through multiple stakeholders and gatekeepers. Healthcare providers were often described as conservative and technology averse. But as the adoption of technology increased outside traditional healthcare systems, it has pushed traditional healthcare system to adapt and adopt these technologies.

Hence companies are finding that healthcare systems are becoming great partners for clinical development, and it is now quite common for these structures to have innovation offices that not only help integrate innovation, but also match and facilitate discussions between innovators and potential end-users. The increased pressure to perform that healthcare facilities now carry has led them to increase their innovation and integration strategies and is a boon for entrepreneurs in many sectors of activity.

13.6 How Has COVID-19 Shifted Investment Perspectives?

The Global Innovation Index[3] shows that investments in innovation have held up well during the COVID-19 crisis and have even reached new heights in certain sectors and regions. Before the pandemic, investment in innovation was at the highest level on record, and research and development spending increased by 8.5% in 2019. When the pandemic hit, people wondered what impact it would have on innovation. Based on past observations, innovation investments were expected to be hit hard. The main indicators of investment in innovation – scientific output, research and development expenditure, filing of patent applications and venture capital operations – nevertheless continued to increase throughout 2020. These indicators show that governments and businesses increasingly recognize that novelty in ideas, products and services will be essential to post-epidemic recovery and growth. However, more data will be needed before a full assessment can be made.

Contrary to the 2008–2009 crisis when the amounts raised and invested had collapsed, activity at the beginning of the year 2020 remained very sustained. Private equity firms weathered the economic cataclysm of COVID-19 by continuing to fund businesses.

A study was carried out by Mazar[4], which surveyed private equity firms and investors to understand the challenges, level of optimism and response strategies in the wake of the COVID-19 crisis. The survey results confirmed that times are tough for private equity players. Deal flow has been reduced, lockdown restrictions made it difficult to close deals and there has been increased demand for support from holding companies.

However, despite these significant challenges, the survey saw a certain level of optimism and resilience from the private equity community and anticipating the materialization of transactions in quality assets:

- 74% of respondents remain on the lookout for new opportunities in the immediate future. Larger funds are the most optimistic about the impact of the crisis on their holdings.
- 79% of respondents said their plan to dispose of portfolio companies will be delayed.
- Despite the monetary impact of the pandemic, 44% of funds have yet to see an increase in investment opportunities in struggling companies.
- 82% of respondents anticipate a U-shaped exit from the crisis.

 **Philippe weights in the investment perspective:** Through my experience, there is a kind of wait-and-see attitude on the part of the "investment" ecosystem. Fundraising deadlines are longer and more complicated. As mentioned previously, sound investment files always come out on top of the fray, with management teams that deliver, a news flow that is well paced and cash management that is particularly well handled.

It seems to me that from this crisis, the strategy of federating several funds in a seed round will be accentuated to be able to finance companies until they reach a milestone of creation of value sufficient to create an opportunity.

13.7 How Has COVID-19 Shifted Recruitment and HR Practices?

This is one sector of activity where we have both noticed deep change as the two years of the COVID-19 pandemic stand as a buffer in the before and after world of hiring trends.

Before COVID-19, finding employees could be seen as a given, and was often taken for granted without a second through. Companies could add a hiring strategy in their business plan or their pitch deck, using superficial terms. Something akin to "By the second year, we will hire two technicians to handle analysis, and we will double the number of lab technicians by the end of the third year to allow us to respond to client requests in 24 hours" passed the reality test. Having a high-level idea of the number of staff to hire, their expected tasks and an actionable hiring plan was enough.

Since COVID-19, this has brutally changed, and having staff lined up and ready to go can demonstrate seriousness of a project, even going as far as being a distinguish factor when raiding funds. For example, having staff part-time ready to ramp up to full time, a university professor ready to join your team as senior scientist or participating in university coop programs can go a long way in highlighting the operating plan in hiring and retaining resources.

There are several hypotheses as to why we are having these challenges. Some believe that COVID-19 accelerated retirement plans for many, reducing the available workforce. Other believe that shift to virtual workplaces has made it harder to "bring back" employees. Finally, there are some that believe there is a generational gap, where the new generation is less inclined to follow the corporate processes of old (the nine to five, working from the office, unpaid overtime and so on).

Whatever the reason, this means that new entrepreneurs must account for a restricted workforce and plan the hiring and retention processes much more carefully than before.

13.8 How COVID-19 Has Shifted This Book

As mentioned earlier, this book was first started during the last phase of COVID-19 in Europe. While North America still had several restrictions in place (restricted movements, vaccine passports and limited interactions between individuals), Europe was slowly re-opening. Some ideas and concepts that we were thinking of adding to the book made complete sense back then, especially when we discussed how COVID-19 shifted our world. Today, these concepts seem horribly archaic, and they are less than 12 months old... We had in our chapter notes concepts such as "COVID-19 has had a durable impact on travel and should limits travel and movement" and "There is less emphasis on meeting face-to-face", all of which seem quite irrelevant when we look at the world today.

But it does contextualize something. While our world changed, one of the major shifts has been the velocity at which things change. As such, we have noticed that these rapid shifts have transpired throughout decision systems, from consumer purchasing habit to corporate decision making, while severely hampering our ability to build models around long-term benefits and slower return-on-investment cycles. If there is something you must remember, it is that right now, there is a strong push towards business models that emphasize short term gains. While this might not mean developing a profitable model within 12 months, it might mean your roadmap will need to quickly include commercialization milestones such as "My First Client" and "My First 100K of sales" much more quickly than we used too.

Notes

1. Pifer R. The shifting digital health investment landscape in 2022, Healthcaredive. https://www.healthcaredive.com/news/digital-health-VC-investment-landscape-2022/617063/ (Last visited 18 May 2023).

2. Gaviria M. and Kilic B. A network analysis of COVID-19 mRNA vaccine patents. *Nat Biotechnol* 39, 546–548 (2021). https://doi.org/10.1038/s41587-021-00912-9 (Last visited 18 May 2023).
3. The Global Innovation Index (GII) is published by WIPO, in partnership with the Portulans Institute and with the assistance of the National Confederation of Brazilian Industry (CNI), the Confederation of Indian Industries (CII), Ecopetrol (Colombia) and the Assembly of Turkish Exporters. https://www.globalinnovationindex.org/Home (Last visited 18 May 2023).
4. Pithois S. COVID-19 and the world of private equity. Mazars Global Financial Advisory Services. https://www.mazars.fr/mazarspage/download/996921/52079801/file/Covid-19+and+the+world+of+private+equity.pdf (Last visited 18 May 2023).

Chapter 14

Challenges in Commercialization for Life Sciences Innovations

---

  As our discussions progressed, we would keep finding challenges to commercialization for companies in life sciences. When it made sense, we integrated them in their relevant section, but for a few, their diversity made them quite eclectic. Hence, we decided to take a simpler approach and introduce them in a consolidated chapter. We will be giving our points of view on commercialization challenges related to digital health, quick thoughts on the role of regulatory frameworks, some thoughts on social media and on AI and life sciences.

14.1 Commercialization Challenges for Digital Health Transformation

Healthcare digital technologies face an uphill battle: not only do they have the traditional challenges of healthcare products since they must demonstrate clinical efficacy, but they also operate in an environment which is notoriously resistant to change and slow to adopt new technologies. Resistance can come from multiple vectors, from the

 DOI: 10.4324/9781003381976-14

processes in place, to the personnel themselves. This resistance is itself understandable: after all, a new technology that fails to deliver not only fails its intended purpose, but also puts lives at risk. The stakes are high, and as such, systems favor minimizing risks. From a more practical perspective, there are four challenges that are often brought up when attempting to commercialize digital health technologies. These are the ownership of data, the integration of technology into legacy systems, the resistance from "older patients" and the resistance from healthcare employees.

The **ownership of data** generated by connected devices is becoming one of the primary contention points during implementation. Questions such as "Who owns this data?" or "Who will do what with it?" must all be answered prior to pitching. On one hand, entrepreneurs want access to as much data as possible, so they can develop further and optimize their technology, potentially developing artificial intelligence agents to analyze the data (and then build prediction or diagnostic tools). In some rare cases, entrepreneur wants to be allowed to sell anonymized data sets that their tools collect to third parties, as this is part of their commercialization model. On the other end, patients want the data to be used mainly for matters related to their health and have little interest (and little incentive) to let private companies have access to its data. With the number of breaches in security related to companies that have stored consumer data, security is fast becoming a key evaluation factor for end-users. To address this, entrepreneurs must address data concerns from all angles, showcasing which data sets they wish to retain for themselves, how they intend to use it and how they plan to secure it.

Philippe weighs on AI commercialization: In recent years many innovative companies have developed technologies related to the processing of data in life sciences. To validate the algorithms, it is essential to reach a critical size in terms of data collected by users or patients. And in disruptive technologies, the essential point lies in accessing or generating data in a context that is of interest to the market. To take the example of neurotechnology, a lot of data exists historically but very little is linked to contexts that interest or will interest investors. While the electrocardiogram (ECG) has become really democratized, the electroencephalogram (EEG) is not yet democratized. Most of the data available are those in the context of analysis in neuropathology or understanding of processes such as sleep. Very little data is available on monitoring in contexts other than medical. It is in progress but the company that will win the race will be the one who has enough contextual or quality data to finally validate their artificial intelligence or feed their self-learning or machine-learning approaches. The best analysis tools are relevant only if they are validated on a (large) representative and high-quality sample (relevance of the data). One example in our ecosystem is the artificial intelligence Watson, created by IBM, which had disappointed because the health data used was not ideal (or available).

Also, entrepreneurs developing digital tools that will be used in clinical settings might uncover that the organizations they are partnering with utilize **legacy systems** to power some component of its information technology infrastructure. These components, often out-of-date or poorly documented, can make a straightforward integration a nightmare. In one case Jean-François worked on, a client had convinced a hospital rehabilitation department to use its app for remote patient monitoring. The company had agreed to pay for all costs related to the projects (secured iPads for the doctors, monitoring devices for the patients), but was stopped by the IT department at the hospital, which, running a system on an old IT infrastructure, could not figure how to securely transfer information from the devices to the electronic health record (EHR). In this case, technology integration was the main barrier to commercialization.

JF weighs in on technology commercialization: There are some strategies that you can employ to address these concerns, and the best one I have found is that you should avoid working on single-product solutions, but rather develop platform solutions.

A single product solution is one that targets a single indication or problem while a platform addresses (or has the potential to address) several indications or problems. To illustrate this, a single-product solution might be an app that is designed to diagnose a single type of cardiac disease, while a platform might be able to address multiple cardiac indications. As such, platforms are usually designed to be scalable and grow as markets evolve. In my experience, hospital IT departments don't have much interest in single-product solutions as the implementation effort and cost to make sure is secure is prohibitive, making it harder to justify the effort. Meanwhile, platform solutions, addressing multiple issues, are easier to justify the effort.

Low levels of computer literacy in **older populations** have also traditionally been a major roadblock that limit the use of some digital health technologies, but this factor is slowly becoming less important as populations get more educated on technology. For example, the 2023 AARP Annual Member survey found that overall use of home safety devices among 50-plus consumers, particularly those used for health or medical purposes, is comparable to that of younger consumers. Furthermore, 21% of the 50-plus use, or are interested in using, technology (e.g., apps) to help them improve their health and well-being, compared to 32% of those aged 18 to 49: more interestingly, that numbers rise to 28% when considering only those between 50 and 59 (compared to older segments, like those over 70+, where interest logs in at 15%).[1] As we go forward, the age gap for technology should slowly disappear, to a point where age is not a primary concern.

For those currently commercializing technologies, waiting until the population catches up to technology usage might not be an option. To address these resistance factors in digital technology, specific strategies and tools must be included in your development for the target audiences that will be utilizing your technologies: if you target a general older public, then developing training tools such as videos can go a long way into increasing the adoption of your technology. Designing simpler user interfaces and limiting the number of disrupting upgrades can also go a long way in increasing adoption and retention.

The last topic we wanted to touch upon is resistance to change by healthcare personnel. In this case, we have seen both sides of the coin: In some cases, personnel were excited about innovation to the point they would openly support and encourage integration, suggesting improvements to increase efficiency. In other cases, personnel would drag their feet to the point the technology would be left unused on a shelf. What we found is that products that had correctly integrated user feedback in their design were more likely to be adopted, as these devices not only answered a pain point but also integrated seamlessly into existing processes. Products that require a lot of change, or that create more extra work are more likely to encounter resistance.

14.2 Regulatory Frameworks

Entrepreneurs in life sciences often operate in a crowded marketplace. They often have a high number of competing technologies (either direct or indirect) and must be very creative about getting customers. This is not something that is exclusive to companies in this space.

Unfortunately, they also operate in one of the most regulated industries in terms of advertising and consumer outreach. They must respect state/provincial and federal regulations. Once a company goes global, regulatory complexity increases substantially. As such, how you address your clients, your claims you are allowed to make as well as the advertising and marketing materials you prepare are all likely covered by many regulatory constraints. In the US, you might be subject to FDA regulatory framework. In Europe, it is the European Food Safety Authority (EFSA).

This regulatory framework is especially surprising for entrepreneurs who have already operated in a less regulated environment and are expanding into healthcare. This happens mostly for those designing apps and software but can also impact those moving from the food to nutraceutical space, for example.

Beyond the government regulatory framework, you might also be subject to industry associations such as PhRMA, which set boundaries on how reps can interact with customers and doctors. Many common strategies and tactics that are quite common in other industries are heavily frowned upon or totally forbidden in this space (such as advertising directly to consumers).

All in all, the important takeaway is that if you are entering this space, take a moment to learn the rules and regulations that will impact your organization.

14.3 Social Media and Life Science Marketers

Stakeholders in the space of healthcare, from medical researchers to practitioners to patients, are looking to make the best decision possible when it comes to selecting the best product or service. Health occupies a specific space when it comes to decision-making, as people are more likely to choose the best solution (best brand, best reputation, best efficacy) even if it is more expensive: people seldom cheapen out when their own health is on the line.

Overworked healthcare systems, the rising growth of self-care, disrupting supply chains have all made stakeholders a lot more interested in researching and documenting the product they will select. From a researcher performing research in a university lab to a parent taking care of their child, quality is fast becoming a key element when selecting a product. While many sources of information exist, social media is a key component in purchasing decisions due to its easy availability, perceived trust and quick updates. We have worked on multiple projects on purchasing processes with everything from University Labs to hospitals, to patients, and while some elements may change from one product to the next, "searching on the internet" is always one of the top two sources during decision-making.

Unfortunately, the pandemic we just lived through has considerably shifted the information landscape. While a few years back, we could trust and rely on major sources of information, the spread of misinformation and the way information is truncated have made the online landscape a mess. It's hard to know what to trust, and companies that adventure in advertising and informing consumers not only have to make sure they have the right message, but they also must be ready to manage the backlash and misinformation that might be associated with their product or service.

Hence it is important to create content that is

- **factually accurate:** you must make sure your publications are well-referenced, have an authoritative tone and look professional. Don't cheapen out on the look'n'feel of your product, as you are being benchmarked against other professional sources; being truthful is another key element, one lie and your credibility goes up in smoke.
- **engaging content:** the dilemma is to create friendly content, that doesn't come out to patronizing or condescending. Remember that stakeholders are constantly being told that they are wrong, and they should be doing this instead: adding your voice to that chorus is the best way to get drowned out. Instead, try positioning your content as approachable, like a guide exploring multiple options.

- **timely**: create content on topics that are current, and of interest. Try to bring parallels between your content and some ongoing events of importance.
- **complies to existing regulations**: as we discussed in Section 14.2, compliance is especially challenging in this place. Having specialized resources to prepare your content is a sound investment to make sure you do not run afoul of existing regulations.

14.4 Challenges of AI and Life Sciences

As we write this section in May 2023, the debate on the use of AI in life sciences is extremely popular, justifiably so. AI is already useful for several research activities, from determining optimal drug delivery mechanisms to repurposing drugs. AI, as a research tool for pharma and biotech, has been used for many years and has demonstrated its value repeatedly.

The next step would logically be to use these tools for market research. Think of all the time you would save if you could just type into an agent "What are my top competitors?" and get a complete list of competitors! Unfortunately, our current experiences have shown that this might be a little premature. To illustrate, we thought we would share an experience Jean-François had using ChatGPT for a market research project.

 **Jean-François weighs in on ChatGPT:** I was preparing some slides for a client, who designs gold nanoparticles. After doing some research, I headed to ChatGPT and asked: "Are there any major pharmaceutical currently developing oncology drugs using gold nanoparticles?" The initial answer was quite interesting.

Yes, there are several big pharmaceutical companies that are currently developing oncology drugs that incorporate gold nanoparticles. Some examples are Pfizer: The company is developing a drug called Aurolux, which incorporates gold nanoparticles and is designed to enhance the effects of radiation therapy in patients with solid tumors.

> *The gold nanoparticles in Aurolux are designed to selectively accumulate in tumor tissue, where they can absorb radiation and produce a burst of electrons that can damage the surrounding cancer cells. Aurolux has shown promising results in preclinical studies, with researchers observing a significant reduction in tumor size and increased survival rates in animal models of cancer. In addition, the drug has been well-tolerated and did not show any significant toxic effects.*

This seemed like valuable information, except for one thing: **Aurolux does not exist.** After searching through the Pfizer website as well as ClinicalTrials.Gov, I could not find any information on Aurolux. As has been demonstrated a few times, ChatGPT had invented an answer out of thin air. As our understanding of ChatGPT grows, it seems that as the AI agent is trying to answer the question, it can, in some instances, pull one out of thin air as a way to fulfill the request.

As new iterations of these tools hit the market, there might come a time when they become a validated tool, but at the time of writing, the use of ChatGPT and others like it for market research seem like asking trouble.

Overall, companies in life sciences must reflect on the role they want to give to AI in their organization. Ideally, they must identify tasks and operations that benefit from automatization, analyze the risk relating to automatizing this process, and then weigh the risk versus additional value generated by using AI in the identified process. This means you will need to identify metrics to evaluate the integration of AI, and be able to quantify it.

Note

1. Kakulla B. *2023 Tech Trends and the 50-Plus*. Washington, DC: AARP Research, January 2023. https://doi.org/10.26419/res.00584.001

Chapter 15

Final Words

 The entrepreneur's journey is an exciting one, which can take many different paths and roads. As a reminder, it starts with the **idea**, passing through the **evaluation of the opportunity**, then **the planning and the strategy**, to materialize and be shared during a **pitch**, to begin a stage of **growth**.

To prepare you in this journey, we have shared with you what we have learned from our experiences and our own journeys. From the entrepreneur side, Jean-François shared the best practices on marketing and how to pitch, while Philippe has thoroughly detailed what happens behind the scenes on the evaluation and due diligence process, sharing some recipes and best practices for success as well as taking you through the art of negotiation. To also prepare you to this important step in your journey, we have also shared what we believe are eight of the biggest mistakes entrepreneurs make when pitching to VCs.

Even if every Entrepreneurship journey starts with **the idea**, don't forget that Business creation is at the same time Darwinian, Brownian and Schumpeterian. Of course, you have to work a lot, but you really have to be lucky! Entrepreneurship is complicated, risky, requires luck, to be resilient, and to adapt. Preparation is the key for success.

Once this idea takes form, you have to **evaluate** it. For an entrepreneur, demonstrating that their innovation is right for the market and will generate interest is a key step to raising funds. As such, it is important to have a framework that lets you validate if your product is a good fit for the

DOI: 10.4324/9781003381976-15

market. One of the frameworks we spent some time explaining is the Product-Market-Fit (PMF). Remember when checking for PMF: Are you evaluating the customer's interest in the solution that you are offering, or are merely investigating a problem/solution fit? In the latter case, you might not necessarily evaluate the interest in the product or service you have developed, identifying a false positive, an interest in the market for a solution to a problem, but not necessarily that your product is the solution people are looking for.

Intellectual property is an essential component of start-ups in life sciences. Your IP portfolio is a necessary part of your story to get the conversation started with investors but be ready to defend your innovation based on results. IP is now being scrutinized a lot more thoroughly than before, and companies to longer get a free pass for "having a lot of IP".

Once you have validated your opportunity, then comes the **planning and strategy step**. By focusing on their customers, startups can create unique products and services that meet customers' needs more effectively than their competitors. A customer-focused approach also enables startups to keep their overhead costs low while still providing value to customers.

Keep in mind that it is common for investors to attend an interesting technology presentation, and start the feedback session by asking the founders:

- **"Who** will use this innovation? **Do they need** it? Are they **looking for it**?"
- **"Who** will pay for this new innovation?"
- "Is it **significantly better** than existing products to justify changing what they already use/already have?"
- "How will this technology **integrate** into existing processes?"

Articulating your business model is essential to demonstrate you are ready for business. A business model is how you plan to structure your commercial operations so you can generate revenues from your innovation (your idea). For this structure to be robust, it is key to make four decisions by defining your commercialization model, your revenue strategy, your corporate strategy and your exit strategy.

An exit strategy is an essential part of your business plan. From the very start of your business venture, you should know how you plan to leave it. For what? Because in fact, exits are when founders and investors get paid!

15.1 Your Idea Is Mature, Your Plan Is Made? Time to Pitch!

To prepare your pitch and the presentation deck, we believe you should be able to address the three following points during your presentation: Ask yourself the right questions such as "How am I going to generate performance for future shareholders (milestones) and give them practical opportunities of benefiting from it (when founders and investors get paid: Exit strategy); gain the support of "ambassadors" and don't neglect the operation side of managing a start-up.

For the presentation deck, we strongly advise you to follow the 10/20/30 rule of presentation proposed by Guy Kawasaki, a Silicon Valley venture capitalist specialized in marketing: "A PowerPoint presentation should have **ten slides**, last no more than **twenty minutes** and contain **no font smaller than thirty points**".

The objective of your pitch is to convince investors that you have an opportunity they cannot miss. But remember that if during your pitch you are thinking "give me money!", the investor will be thinking "give me some performance!".

There are two essential elements to successfully start your investment journey: First, as we mentioned earlier, it is essential to have a clean and precise pitch. Second, it is essential to be able to find the right person to pitch to.

To ensure that you reach the right contact, it is essential to carry out a small market study and to characterize the teams and funds in the sector that you are operating in. The choice of the investor and the right person to pitch to depends on several criteria: The stage of intervention, the status of the private equity firm, the investment amounts granted, sectors of intervention, the geographical coverage as well as their type of stakeholding.

Investors' experience is key. Do not hesitate to contact companies in their portfolios or consult the websites of investment funds, they will have information on their investment policy. You will be able to look at their previous investments and validate their targets (amounts, stage of development, sector etc.). As soon as you have identified the right VC, the one who will have the most affinity with your project, we believe that the best way to approach the person identified is to attend a conference and be able to contact them directly.

How VCs make investment decisions is a very systemic process. It is organized in successive well-established phases. Each of these can be very intensive and long, depending on the actions of either parties (the start-up or the investor). It is particularly important to understand and integrate the deal flow funnel into your approach (sourcing, deal screen, teams review, due diligence, investment committee, negotiation and closing). A typical VC might see a high number of opportunities for his fund (over 1,000), he will usually end up investing in less than 1% of these opportunities.

Even if there is no particular secret sauce to negotiating with a VC, it is important to keep in mind that your journey will start together after the closing and this until an exit (IPO and M&A for the most part). Also, we recommend to always be in a state of mind of 'being side by side' versus 'facing each other' when entering negotiations.

We've had the opportunity to review and interact with hundreds of startups over the years. This allowed us to develop a quick checklist of basic mistakes entrepreneurs make. To avoid these errors, it is good to know them. We've rounded them up into a list of eight classic mistakes life science entrepreneurs make: Underestimating or overestimating their pitch deck, not believing narrative is everything, underestimating competition and waiting to raise funds when needed, overestimating or underestimating the protection or value of their IP, overestimating themselves and underestimating cash flow management and finally neglecting their brand.

Successful fundraising is not a long calm river, especially the first fundraiser. It's a marathon of at least 12 to 24 months. To maximize your chances of convincing investors, remember two things: Preparation and promoting traction (business traction).

Nonetheless, the COVID-19 crisis left its mark: The interventionist policies (injection of capital and aid to companies) during the crisis caused countries' debt levels to explode. In addition, the shortages of raw materials, which had already started before the pandemic, are all the more visible and marked in a turbulent global context. These prevent industries from operating normally and contribute to inflation.

The war in Ukraine, for example, has complicated international relations: This war has greatly disrupted the global geopolitical context. Beyond Ukraine, it is symbolic and involves almost the whole world, especially the great ideological powers; USA, Europe, Russia and China. It is a war that exerts strong pressure on the world economy, and that generates insecurity and uncertainty. This implies greater risk aversion for financiers. In addition, the sanctions put in place against Russia have disrupted the energy supply (the cost of energy has exploded) on which certain European countries are very dependent.

In addition, Russia being a major wheat producer and Ukraine being considered "the breadbasket of Europe", the food supply, and in particular cereals, is also disrupted. Finally, some large groups have decided to cease their activities in Russia or in collaboration with Russia, which inevitably leads to a loss of profits.

The bubble of tech values (startups, innovation, and the digital sector) which had formed in 2020 and 2021 had actually started to burst at the end of 2021 in the United States, and 2022 and the geopolitical context came to accelerate this burst in the world.

In 2022, the Nasdaq had its worst session in two years. Tech stocks have been volatile in this period of uncertainty, and their prices have fallen (Apple; Netflix, Tesla, Amazon etc.) Institutional investors (Limited Partners – LPs) have then left these risky stocks to turn to less volatile assets and become more attractive again following the rise in interest rates.

These changes in the geopolitical, macro-economic and even health contexts lead to upheavals in the behavior of economic factors, including investors, who are forced to adapt.

When it comes to VCs, we can see some major shifts in behavior:

- B2B has an advantage over B2C: Large companies, which are often better protected, will be favored because the risk is lower and the decline in consumer purchasing power is obviously to be taken into account.
- The impact is a major factor, startups that integrate social and environmental dimensions are of greater interest to investors.
- Investors are on longer decision cycles and are more demanding during due diligence, this requirement will be reflected in particular on the study of your financial documents.
- Investors focus more on the turnover/profits that you will be able to generate, when in recent years the market has been looking for hyper growth at all costs.

Thus, it is essential to be prepared during your meetings with investors because they will be even more attentive and demanding during the analysis of your documents and will test the solidity of your project. Keep in mind that the investor seeks to evaluate your qualities and your potential as a founder. The more you are prepared during your exchanges with funders, the more luck you put on your side. Remember that the race for fundraising and capital is very competitive, many are called for few chosen. Anticipation is therefore the key!

So, we do love lists. And so, we end this book with seven questions you should ask yourself before you go out there and pitch.

1. Does my technology address an unmet need?
2. Does my technology display a strong level of technological innovation?
3. Is my technology sufficiently different from competing technology?
4. Is my technology protected by a strong IP portfolio?
5. Have I clearly identified my path to commercialization?
6. Does my team display a capacity to execute my technology, my commercial plan and my IP/regulatory plan?
7. Is my story awesome?

Once you address these seven points, you are well on your way to your first financing.

Index

Note: Locators in *italics* represent figures; **bold** indicate tables in the text and page numbers followed by 'n' refer to notes.

Printed in the United States
by Baker & Taylor Publisher Services